EMERGING MATERIALS FOR CIVIL INFRASTRUCTURE

State of the Art

EDITORS-IN-CHIEF
Roberto A. Lopez-Anido and Tarun R. Naik

EDITORS
Gary T. Fry
David A. Lange
Vistasp M. Karbhari

ASCE American Society of Civil Engineers
1801 ALEXANDER BELL DRIVE
RESTON, VIRGINIA 20191–4400

Library of Congress Cataloging-in-Publication Data

Emerging materials for civil infrastructure : state of the art / editors-in-chief, Roberto A. Lopez-Anido and Tarun R. Naik ; editors, Gary T. Fry, David A. Lange, Vistasp M. Karbhari.

p. cm.

Includes bibliographical references and index.

ISBN 0-7844-0538-7

1. Materials. 2. Building materials. I. Lopez-Anido, Roberto A. II. Naik, Tarun R. III. Fry, Gary T. IV. Lange, David A. V. Karbhari, Vistasp M.

TA403 .E46 2000

624.1'8--dc21 00-063960

Notices

Library of Congress Catalog Card No: 00-063960
ISBN 0-7844-0538-7
Manufactured in the United States of America.

Foreword

This book presents a review of emerging materials for use in civil engineering infrastructure. Emerging materials have attractive material properties, durability, economics, and/or environmental impact. They include novel and new materials, and traditional materials with profound potential in new applications. The materials and material systems may be conceptual and still in research and development; or may be newly introduced whereby adoption by the community is only beginning. A material or materials system is considered to have matured beyond the phase of an "emerging materials" if it is widely used in practice or if it has been represented in well-established design codes or specifications. In that rein, new and dramatically different varieties of established materials, such as aluminum or even masonry, could be considered as belonging to the class of "emerging materials" if their specific characteristics are such that existing specifications and codes either do not apply, or have to be drastically modified for their use.

The Committee on Emerging Materials, under the leadership of Tarun R. Naik since its inception, has promoted activities in the field of emerging materials through collection, assessment, and dissemination of information regarding materials characterization, processing, evaluation, applications, performance, and acceptance for Civil Engineering construction including infrastructure. Furthermore, the Committee is charged with encouraging research, development, demonstration, and education relating to emerging materials. The culminating endeavor of the Committee has been to facilitate this special publication on the state-of-the-art of emerging materials. Information about the Committee activities, including the preparation of this special publication, can be found in the Technical Divisions section of the ASCE web site http://www.asce.org

An Editorial Team of Committee members edited the manuscripts and compiled the book. The Editorial Team organized this publication through a selection of emerging materials to guide their use by practicing engineers to solve civil infrastructure challenges. The book is structured in chapters that address specific classes of materials highlighting the most recent developments in materials technologies relevant to civil infrastructure. The rationale for the book organization is to provide the practicing engineer, student, or a lay reader with an easy reference to a number of material systems that are increasingly being developed and considered for use in civil engineering. Hopefully, the contents will impart both fundamental knowledge and an introduction to these classes of materials, while simultaneously providing a wealth of useful and state-of-the-art information for engineers and scientists already exposed to these materials.

Editors-in-Chief

Roberto A. Lopez-Anido
Tarun R. Naik

Editors

Gary T. Fry
David A. Lange
Vistasp M. Karbhari

Abstract

This special publication presents a review on the state-of-the-art on emerging materials for use in civil engineering infrastructure. Emerging materials include novel and new materials, and traditional materials with profound potential in new applications. For the purposes of differentiation, a material or class of materials, is considered "emerging" if its use has not yet developed to a stage wherein well-established guidelines, codes and specifications exist for its use. This publication was compiled based on a selection of emerging materials to guide their use by practicing engineers to solve civil infrastructure challenges. The book is structured in chapters that address specific classes of materials highlighting the most recent developments in materials technologies relevant to civil infrastructure. The rationale for the book organization is to provide the practicing engineer, student, or a lay reader with an easy reference to a number of construction material systems that are increasingly being developed and considered for use in civil engineering.

Peer Review Policy

Each chapter published in this volume was evaluated through three peer reviews. The first review was by the members of the Emerging Materials Committee, the second review was by the members of the Materials Division Executive Committee, and the third review was by external experts in the field. The authors addressed all of the reviewers' comments.

The quality of the papers in this publication reflects not only the obvious efforts of the authors and the editors, but also the work of these peer reviewers.

Acknowledgements

The Editorial Team acknowledges the dedication and contributions of participating authors, the members of the Emerging Materials Committee, the members of the Materials Division Executive Committee and the external expert reviewers. In particular, they acknowledge the guidance and assistance of John E. Bower who served as the contact member of the Executive Committee.

Emerging Materials Committee

Preface

The Executive Committee of the Materials Engineering Division commends the Emerging Materials Committee on the publication of this book presenting a state-of-the-art picture of new trends in construction materials.

It is appropriate, as the 21st century begins, to review and examine newer materials. These will be the materials that designers and constructors, testing laboratories, and specification groups are introducing and synthesizing into our civil infrastructure, with the help of new research data and new educational presentations.

It is also significant that this book is published now, just as the new Construction Institute of ASCE begins -- an Institute that integrates the Materials Engineering Division and the Construction Division with their broader communities, with missions of promoting excellence in construction and of advocating quality and technological advancement in the application of materials in construction.

It is our expectation that this book can assist in these endeavors.

Materials Engineering Division Executive Committee

Lawrence Bank
John E. Bower
Srinivasa Iyer, Chair
Antonio Nanni

Contents

Craig A. Rogers,[1] Victor Giurgiutiu,[2] and Christopher K. Y. Leung[3]

Chapter 1

Smart Materials for Civil Engineering Applications

Abstract

A review of the emerging field of smart materials and structures, with special emphasis on civil engineering applications is presented. Sensing, actuation, and self-repairing applications are covered. The smart materials used for sensing applications cover optical fibers, piezoelectrics, smart tagged composites, and electro/magnetostrictive materials. The smart materials used for actuation applications include piezoelectric and electrostrictive materials, magnetostrictive materials, and shape memory alloys. The self-repairing materials encompass special varieties of cementitious and polymeric materials. For each smart-materials category, a brief description of the physical principles and theoretical basis is given. Major research results and proof of concept demonstrations are summarily reviewed. The advantages and limitations of each smart materials technology are presented and discussed. The chapter ends, with an overall presentation of the current development status in the smart materials and structures field for civil engineering applications, followed by a brief outline of future research needs.

Keywords

Smart, adaptive, intelligent, fiber optics, Bragg gratings, piezoelectric, smart composites, tagged composites, electrostrictive, magnetostrictive, shape memory alloys, ferroelectric, ferromagnetic, PVDF, self-repairing, health monitoring.

Introduction

Infrastructure decay is a severe problem faced by many developed countries. A major expense in any rehabilitation program is the high cost of structural inspection. Moreover, due to the uncertainties of structural assessment procedures, minor degradations that can be easily repaired may develop into major damages requiring

---

[1]Professor, Mechanical Engineering Department, University of South Carolina, 300 S. Main St., Columbia, SC 29208

[2]Associate Professor, Mechanical Engineering Department, University of South Carolina, 300 S. Main St., Columbia, SC 29208

[3]Associate Professor, Department of Civil Engineering, Hong Kong University of Science and Technology, Clear Water Bay, Kowloon, Hong Kong

major repairs or complete re-construction. To reduce the cost of structural rehabilitation, it is highly desirable to have materials that can provide information on their own condition, or even better, materials with the capability of self-repair.

New structural design concepts also impose new requirements on material behavior. Recently, there have been considerable interests in the use of active control to reduce the vibration and damage of structures subject to dynamic load. To make active control possible, materials with actuating capability need to be developed.

In this chapter, we broadly classified smart materials into the following categories:

(i) sensing materials - materials that can provide information on its current condition or 'health'.

(ii) actuating materials - materials that can change dimensions under external stimulus

(such as heating/cooling or the introduction of an electromagnetic field). When restrained, an actuating force will be provided.

(iii) self-repair materials- materials that can repair themselves, for example, through the automatic healing of cracks.

It should be noted that some materials exhibit more than one of the above properties. It is envisioned that smart materials will be used in combination with conventional materials to construct a smart structural system. This often requires the use of a computer as the 'brain' or decision-maker of the smart system. For example, to construct a smart concrete structure, optical fibers can be embedded at various locations to monitor the dynamic strain. After the strain information is analyzed by a computer, a signal will be sent to a power supply, which will then apply an electric field to an electrostrictive actuator attached to the structure. The force provided by the actuator can then reduce the vibration of the structure.

In the following sections, sensing materials will be described first, followed by actuating and self-repair materials. After each type of material is described, the current limitations will be highlighted and major research challenges will be identified.

Sensing Materials

Optical Fibers

Optical fibers were first developed in the 60's for telecommunication applications. An optical fiber is a glass fiber with a cylindrical core surrounded by a concentric cladding (Figure 1). The refractive index of the cladding is higher than that of the core. When light is sent into the core at a low angle, total internal reflection takes place and the light is transmitted along the fiber with very little loss. With an optical fiber, signals can be transmitted over very long distances without the need for amplification. To protect the optical fiber against scratches and environmental effects, a coating is always applied onto the fiber during the drawing process. In most cases, the coating is made of a polymer (such as acrylate or polyimide). For special applications, metallic coating such as aluminum or gold are employed.

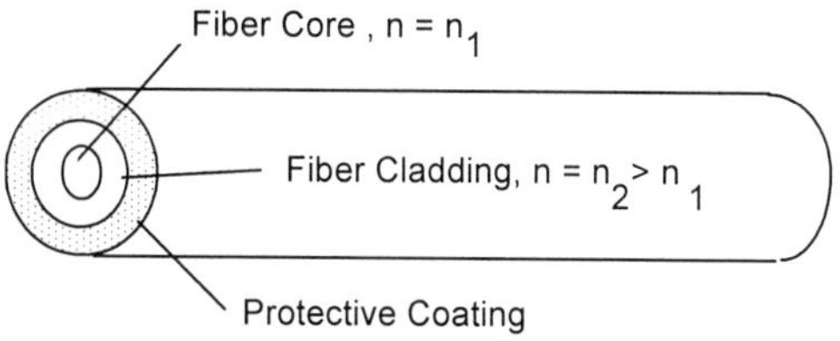

Figure 1 – *Typical optical fiber architecture*

When an optical fiber deforms due to mechanical or thermal effects, the optical signal will be affected. For example, when a fiber is stretched, light travels a longer distance from one end to the other; there will be a change in the phase of the output signal. When a fiber is bent, some of the light will refract from the core into the cladding and get dissipated. As a result, the intensity of guided light will decrease. In fibers with gratings, light of a particular wavelength is preferably reflected. When the fiber is strained, the reflected wavelength is shifted. These changes in light properties form the basis of fiber optic sensing. By coupling an optical fiber to a structure; monitoring the change in light intensity, phase or wavelength at the output, the structural condition can be verified.

During the last two decades, a great deal of work has been carried out in the development of fiber optic sensors for various measurands including distance, velocity, strain, temperature, cracking, pressure, corrosion, maturity of curing, chemical content as well as intensity of high-level electromagnetic fields. Compared with conventional instruments, advantages of fiber optic sensors include:

(i) low signal loss, thereby being suitable for remote sensing,

(ii) immunity to low-level electromagnetic noises common in sensing applications,

(iii) not vulnerable to damage by lightning, which is important to structures such as bridges or dams,

(iv) small size and light weight; hence unobstructive when installed in a structure,

(v) possibility of multiplexing, i.e., having a number of sensing points along a single fiber,

(vi) feasibility of distributed sensing. Any point along the fiber can act as a sensor. This property is critical for crack sensing in concrete structures.

Fiber optic sensing is now a well-established research field with thousands of papers already published in the literature. It is not possible to provide a complete review of the field in this short chapter. We will therefore focus on several topics of particular interest to civil engineers, namely the sensing of strain, cracking and moisture content. For a more general coverage of fiber optic sensing, especially its application in other areas, the reader can refer to the book by Culshaw (1996), as well as the various proceedings on Smart Structures and Materials published by SPIE.

Strain Sensing with Optical Fiber – Fiber optic strain sensing is most often carried out with interferometric techniques. To explain the technique, it is best to consider the Mach-Zehnder interferometer that consists of two fibers. A light beam is split and coupled into two separate fibers. One fiber is attached to a structure while the other is left free. At the other end, signals from the two fibers are re-combined and monitored. When the structure is loaded, the length of the attached fiber will change, leading to an increase in optical path for the light. The corresponding phase shift will lead to a periodic variation of light intensity when light signals from the two fibers are re-combined. Each fringe (or complete periodic variation of light intensity) corresponds to a change in length equal to the wavelength of light, which is around one micron. Counting the number of fringes can provide enough resolution for the measurement of relatively large strains. For small strain measurements, feedback loops have to be employed to convert the periodic intensity into a linear variation in voltage output (Ezekiel, 1992).

The Mach-Zehnder interferometer has several disadvantages. Two fibers need to be used at each sensing point. Very tight control of bonded fiber length is required to ensure a consistent gauge length, so the measured length change can be correctly translated into a strain value. Also, when low-cost light sources such as LED are employed, the low coherence length requires close matching of the length of the attached and free fibers. In the instrumentation of large structures, long fiber lengths are required and close length matching can be very difficult.

To overcome the above shortcomings, a Fabry-Perot sensor can be employed. An example of Fabry-Perot sensor (called the extrinsic Fabry-Perot Interferometer, EFPI) is shown in Fig. 2. The sensor consists of two optical fibers with partially mirrored ends held at a distance apart by a small glass tube. When light signal is sent into the fiber, part of it will reflect at the first glass/air interface, while another part will reflect at the second interface. The two reflected signals then interfere with one another. On loading, the air gap will change in dimension, causing a change in the difference in optical path of the two light beams. With this configuration, a single fiber can be used at each sensing point. The gauge length is fixed when the EFPI is manufactured. The distances traveled by the two light beams differ only slightly; are much smaller than the coherence length of any laser. The application of EFPI for strain sensing has been demonstrated in various laboratory and site studies. Masri et al (1994) use EFPI for cyclic strain monitoring in a concrete structure and show that the signal/noise ratio is much better than that of strain foil gauges. Habel (1994, 1995) and co-workers have applied EFPI to strain monitoring of concrete during the curing and service stages, as well as strain monitoring of prestressing tendons in a concrete bridge.

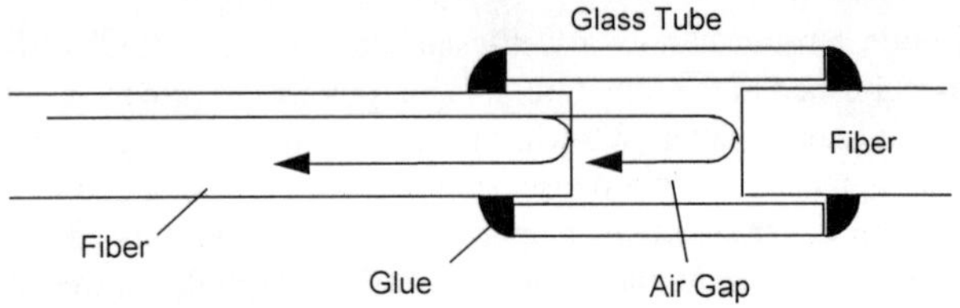

Figure 2 – *The Extrinsic Fabry-Perot Interferometer*

The interferometric technique described above measures the change in phase relative to a certain state. If the signal is turned off for some reasons, the reference point will be lost, as it is impossible to determine the number of fringes the output has gone through during the 'down' time. To counter this drawback, Tran et al (1993) developed an absolute extrinsic Fabry-Perot interferometer (AEFPI), which employs white light interferometry to obtain the absolute size of the air gap. The applicability of such a sensor has been demonstrated through the monitoring of strain in a prestressed tendon (de Vries et al, 1995).

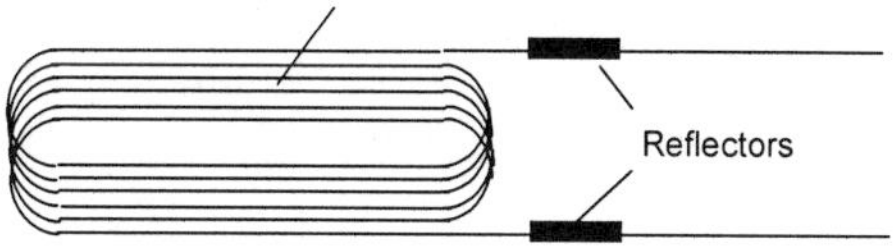

Figure 3 – *A Large Strain Gauge made of Optical Fiber Loops*

For absolute strain measurement, the most direct technique is to measure time for light to travel over a certain distance; how this time value changes as the fiber is strained. Such a technique has been employed by Lou et al (1995) for strain monitoring in a composite jacket for column retrofitting. They developed a sensor that consists of many loops of fiber glued to the structure (Fig. 3). At the two ends of the loops are two optical reflectors. By monitoring the Fresnel reflection from the two reflectors, the total length of fiber in the loops can be calculated from the times at which the reflections reach the detector and the light velocity in the fiber. As the fiber is stretched, the new length and hence the strain can be calculated from the new time values. This method has the advantage of directness and simplicity in data interpretation. Also, one can easily obtain strain at different parts of a structure by looping different portions of the fiber at the appropriate locations. Its major disadvantage is the need of a very short pulse to obtain the required time resolution. This makes the required opto-electronics system very expensive. Also, a long length of fiber is required between the reflectors to provide a high strain resolution. This limits the application of the sensor to the measurement of strain over relatively large gauge lengths.

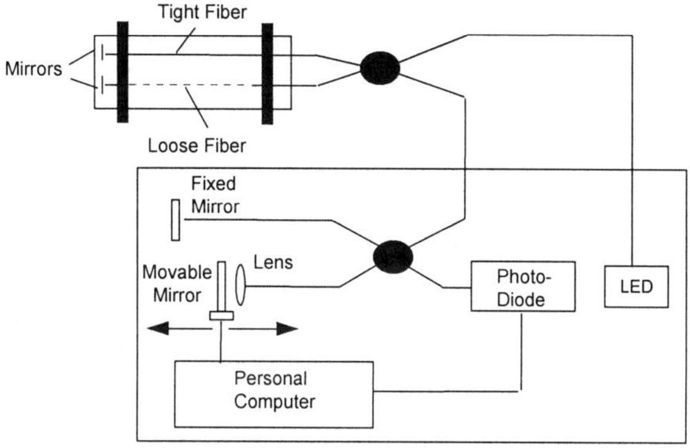

Figure 4 – *Low Coherence Interferometer for Absolute Strain Measurement*

Another set-up for absolute strain measurement has been developed by Inaudi et al (1994). The sensing principle, which is based on white-light interferometry, is illustrated in Fig. 4. In this case, the sensor consists of two fibers gripped by a pair of 'wings' that can be coupled to a structure. One fiber is tight while the other is loose. On loading, only the tight fiber will be stretched and a light path difference is resulted between the two fibers. The signal demodulator consists of a pair of fibers, one terminating at a fixed mirror and the other terminated in front of a movable mirror. With the set-up in Fig. 4, the photo-diode will register a signal only if the path difference between the fixed and movable mirror is the same as that between the tight and loose fibers. The absolute strain can hence be obtained from the position of the movable mirror. Such a sensor has been applied to the deformation monitoring of concrete bridges (Inaudi et al, 1996) as well as underground excavations (Inaudi et al, 1995).

Recently, there is also considerable interest in the development of absolute strain sensors based on the Bragg grating. A Bragg grating is a periodic variation of refractive index along the length of the fiber. It is normally created by interfering two powerful laser beams shooting onto the side of the fiber. The grating period can be controlled by varying the laser beam angle. When a broad band signal passes through the optical fiber, a wavelength corresponding to the period of index variation will be preferentially reflected. When the fiber is loaded, the index variation period will change, leading to a shift in the reflected wavelength. When strain values are required at more than one point along the fiber, gratings of different periods can be created. The reflected signal will then exhibit a number of peaks corresponding to each grating. By monitoring the shifts of all the peaks, strain at each point can be deduced. Cost-effective systems for the determination of shift at multiple sensing points have been developed separately by Alavie et al (1993) and Kersey *et al.* (1993). Field trials of multiple point dynamic strain sensing with Bragg gratings have been carried out in Calgary, Canada (Maaskant *et al.*, 1993) and New Mexico (Davis *et al.*, 1996).

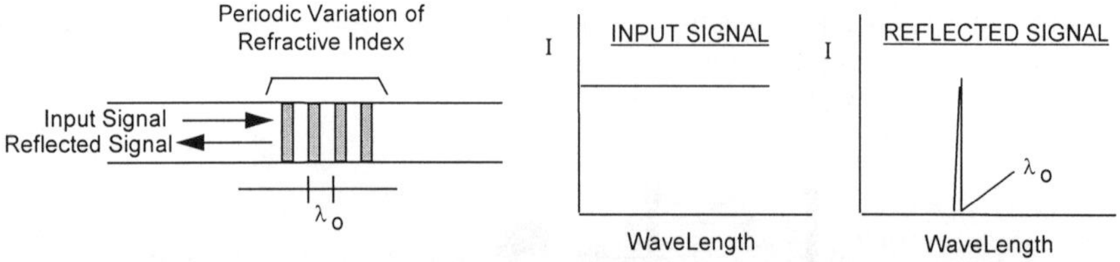

Figure 5 – *Principle of Strain Sensing with the Bragg Grating*

In their recent work, Measures (1997) and Volanthen et al (1997) have independently demonstrated the possibility of distributed strain sensing with Bragg sensors. The continuous strain distribution over a grating (as much as 10 cm in length) can be determined. This capability will make possible the accurate measurement of strain gradients in structural components. Examples include stress at the anchorage of prestressing tendons, stress around bolt holes and stress at bonded connections.

Crack Monitoring with Optical Fibers – In many structures, failure is preceded by the formation of cracks. For concrete structures, small cracks may not affect structural integrity, but they may allow the penetration of water and salt yielding long term durability problems. Crack monitoring is therefore an important aspect in the health monitoring of structures.

The early fiber optic crack sensors (Rossi, 1989, Huston et al, 1992) rely on the breakage of a fiber to produce a sudden drop in transmitted optical power (or, alternatively, a sudden increase in reflected power). With a single fiber, it is not possible to monitor crack growth. Hale (1992) designed a sensor with several fibers. Crack propagation perpendicular to the fiber direction can then be monitored, as breakage of each fiber will lead to partial loss of the forward signal. The major disadvantages of these sensors lie in the difficulty in controlling the failure consistency.

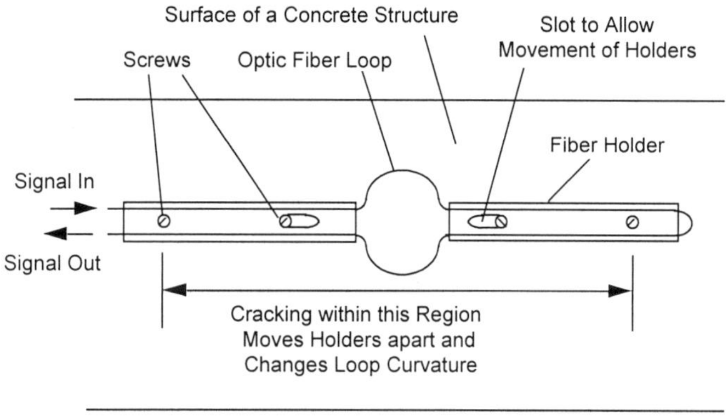

Figure 6 – *The Crack Sensor developed by Wolff and Messelier*

To achieve consistency, all fibers need to be coupled to the structure in exactly the same way - a task not easily accomplished on site. In addition, since glass is a brittle material with high material variability, even fibers bonded in the same way may fail under very different strains.

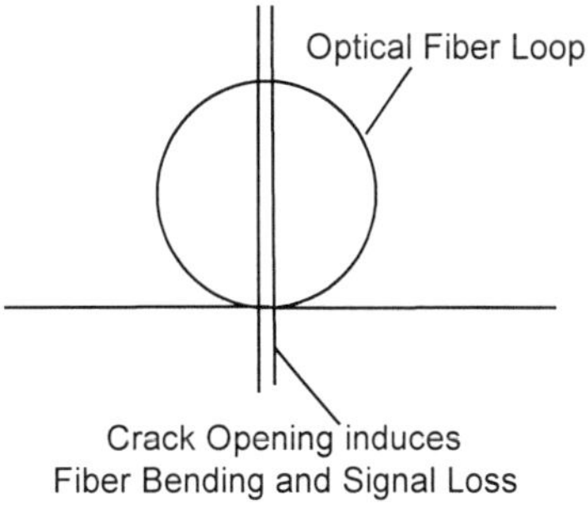

Figure 7 – Ansari's Crack Sensor

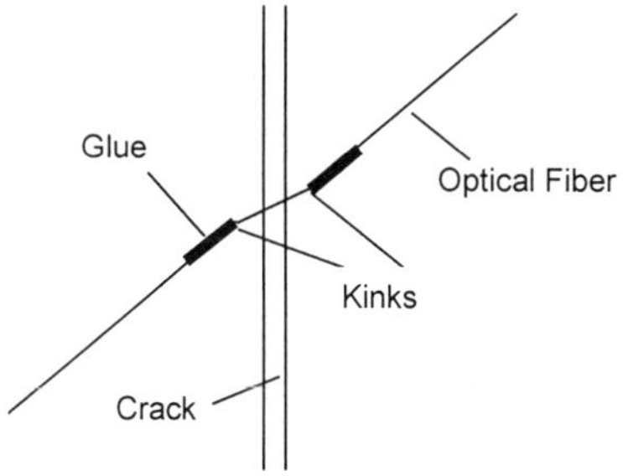

Figure 8 – Boss and Wanser's Crack Sensor

Fiber optic sensors with crack monitoring capability have been developed by Wolff and Messelier (1992), Ansari et al (1993); Voss and Wanser (1994). Wolff and Meisseler's crack sensor (Fig. 6) is an optical fiber sensing system attached to the exterior surface of concrete structures for the monitoring of displacement between the points of attachment. If the attachment points are too close, a crack may not pass through the sensing system and if the points are too far apart, sensitivity will be low and it is not possible to distinguish between the presence of one widely opened crack or several narrower ones. The sensor developed by Ansari et al (Fig. 7) is a 'point' sensor, which means that it can detect and monitor the opening of a crack only if the crack passes through a small sensor loop 31.75 mm in size. Moreover, in order to provide quantitative information on the crack opening, the crack has to pass exactly through the center of the loop because this is the condition under which the loop is calibrated. Thus, while the sensor is useful in experimental fracture mechanics, where crack location can be controlled, its application to real concrete structures is very limited. Voss and Wanser's sensor (Fig. 8) also relies on cracking to occur between two points where the fiber is glued onto the structure. For large or small spacing between these two points, this sensor exhibits the corresponding limitations exhibited by the two sensors described above.

A new crack sensing concept that does not require a-priori knowledge of crack location was recently proposed by Leung and co-workers (1997). The sensor principle is illustrated in Fig. 9. While it is not possible to predict exact crack locations, the crack opening direction can be accurately determined from analysis. Once the crack direction is known, an optical fiber can be coupled to the structure in such a way that it is making an angle to the crack. One approach, illustrated in Fig. 9, is to couple a 'zigzag' fiber to the structure. For easy installation and proper protection of the sensor, the fiber is first embedded into a polymeric sheet that is then glued to the structure. Light is sent into the fiber from one of its ends; the back-scattered signal (at the same end) is monitored as a function of time.

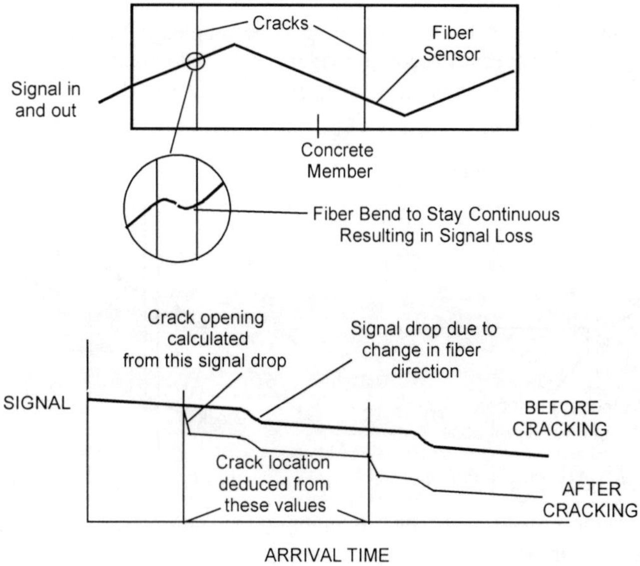

Figure 9 – Distributed Crack Sensing Concept proposed by Leung and co-workers

This technique of signal monitoring is known as optical time domain reflectometry or OTDR. Before any crack is formed, the plot of signal versus time is essentially flat, showing drops only at the turning points of the zigzag fibers. (Note: these losses are shown for a general case - it is possible to use a small curvature at the turning point to avoid the loss). Once cracking occurs, the optical fiber, which is at an angle to the crack, needs to bend to stay continuous. Bending will cause light loss from the fiber core to the surrounding. As a result, the back-scattered signal will show a sudden drop. Knowing the light velocity inside the fiber, the location of the crack can be calculated from the time at which the loss occurs. Also, from the magnitude of the loss, the crack opening can be deduced if a 'calibration' relation is available. Preliminary experimental results have shown that small crack openings (about 0.1 mm) can be detected with the proposed technique. Optimization of sensor design is currently under way to achieve a compromise between sensitivity (increased with loss at each crack) and maximum number of detectable cracks (limited by the total loss budget of the OTDR system).

Water Detection with Optical Fibers – The capability to detect water content is useful in many civil engineering applications. For example, leakage from pipelines can be detected. By knowing the water content in soil masses and soil slopes, the load bearing capability and failure probability can be assessed. Also, in applications involving grouting, the sensing of water content can provide information on location of dry spots where the grout fails to reach (Michie et al, 1994).

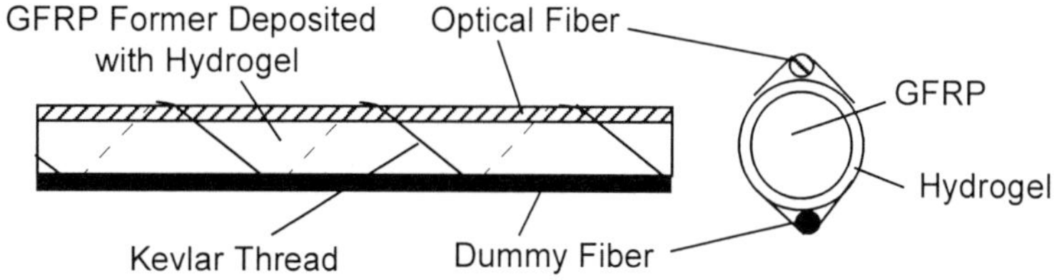

Figure 10 – An Optical Fiber Sensor for Water Content

A sensor for moisture detection, developed by Michie et al (1994) is shown schematically in Fig. 10. The sensor consists of a glass fiber reinforced plastic (GFRP) with a layer of hydrogel deposited on the surface. On top of the hydrogel are an optical fiber and a dummy fiber, both tightly held by a helically wound Kevlar thread at 2 mm pitch. As the gel absorbs water and expand, the optical fiber will be pushed onto the thread, resulting in bending and signal loss. By monitoring the loss along the fiber with OTDR, distributed sensing of moisture content can be carried out.

Advantages and limitations of Fiber Optics Sensors – The advantages of optical fibers over conventional transducers have already been discussed in section 2.1 and will not be repeated here. At the present, the widespread use of the technology is restricted by a number of limitations. The cost of signal demodulation systems is still high. With advances in optoelectronics; innovative developments of more effective demodulation schemes, the cost can be expected to decrease. While the applicability of many sensing concepts (for strain, cracks, moisture etc) have been successfully demonstrated in the laboratory, there is still little experience with application of optical fiber sensors in large-scale civil structures. The long term durability and stability of the sensing system in the real world environment is not clear. When multiple sensors are required (which is the norm for civil structures), the maximum number of sensors that can be economically multiplexed together needs to be considered. In addition, in a large structure, information will be required at thousands of points. The development of innovative distributed sensing techniques is therefore an important area for future research.

Piezoelectric Materials

Piezoelectricity describes the phenomenon of generating an electric charge in a material when subjecting it to a mechanical stress (direct effect); conversely, generating a mechanical strain in response to an applied electric field. Piezoelectric properties occur naturally in some crystalline materials and can be induced in other polycrystalline materials. The distortion of the crystal domains produces the piezoelectric effect. The domains may be aligned (poled) by the application of a large, field, usually at high temperature. Subsequent application of the electric field will produce additive strains in the local domains, which translate into a global strain in the material. The piezoelectric effect was discovered in 1880 by Pierre and Jacques Curie. The direct piezoelectric effect has been used for a long time in sensors such as accelerometers. Use of the converse effect, however, has until recently been restricted to ultrasonic transducers. Recent developments in piezo-electric ceramics have spurred the use of piezoelectric materials in a variety of sensor and actuator applications. Piezoceramics are compact, have a very good frequency response and can be easily incorporated into structural systems. Within the linear range, piezoelectric materials produce strains that are proportional to the applied electric field or voltage. Induced strains in excess of 1000 μstrain (0.1%) have recently been reported. These features make these materials very attractive for a variety of applications.

Principles of Piezoelectricity and Types of Piezo Materials – There are several approaches to modeling the constitutive behavior of piezoelectric materials. The first approach is that taken by material scientists. In this model, the effects of the microstructure and the chemical composition of the material properties are studied. This approach is useful in improving the basic electromechanical properties of the piezoelectric materials; however, the parameters involved in such a model are not in a form where they can be used directly by structural engineers in structural equations.

The second approach is an electrical one, where an equivalent electrical circuit is developed for the piezoelectric material. This is useful for representing the effect of the piezoelectric material on the external electrical circuit. This approach does not deal directly with the mechanically induced strains, which are the source of control in structural applications. Nevertheless, this modeling is important to structural control engineers because it is through this model that the electrical power required to drive the piezoelectric actuators/sensors bonded to the structure is computed.

The third approach is the thermodynamic approach, which is all encompassing and can deal with all forms of energy and their inter-conversion. Thermodynamics is concerned with restrictions on the responses of bodies arising from the energy balance and the entropy inequality. Without these restrictions, mathematically characterized ideal bodies may not represent physical bodies and their predicted response may not be real. This approach, although best, can be unnecessarily complicated for use in the solution of typical structural problems.

The electro-mechanical approach is the fourth approach and the one that is the most popular with design engineers. It provides the relationship between the electrical and mechanical variables in a manner that can be incorporated into the existing constitutive relations of typical isotropic or orthotropic structural materials. The details of this approach are presented herein.

For linear piezoelectric materials, the interaction between the electrical and mechanical variables can be described by linear relations of the form. The general constitutive equations of linear piezo-electric material behavior, given by ANSI/IEEE Standard 176-1987, describe a tensorial relation between mechanical and electrical variables (mechanical strain S_{ij}, mechanical stress T_{ij}, electrical field E_i; electrical displacement D_i) in the form:

$$\begin{aligned} S_{ij} &= s^{D}_{ijkl} T_{kl} + g_{kij} D_k \\ E_j &= -g_{jkl} T_{kl} + \beta^{T}_{jk} D_k , \end{aligned} \tag{1}$$

where s^{D}_{ijkl} is the mechanical compliance of the material measured at zero electric displacement ($D = 0$), β^{T}_{jk} is the dielectric impermittivity measured at zero mechanical stress ($T = 0$); g_{kij} is the piezoelectric voltage coefficient that couples the electrical and mechanical variables. The second equation reflects the *direct piezoelectric effect*, while the first equation refers to the *converse piezoelectric effect*.

Another formulation of these equations outlines the current produced by the piezoelectric effect, i.e.,

$$\begin{aligned} S_{ij} &= s^{E}_{ijkl} T_{kl} + d_{kij} E_k \\ D_j &= d_{jkl} T_{kl} + \varepsilon^{T}_{jk} E_k , \end{aligned} \tag{2}$$

where s^{E}_{ijkl} is the mechanical compliance of the material measured at zero electric field ($E = 0$), ε^{T}_{jk} is the dielectric impermittivity measured at zero mechanical stress ($T = 0$); d_{kij} is the piezoelectric displacement coefficient that couples the electrical and mechanical variables. Again, the second equation reflects the *direct piezoelectric effect*, while the first equation refers to the *converse piezoelectric effect*.

The mechanical variables in the foregoing equations are the stress T_{ij} and the strain S_{ij}, while the electrical variables are the electric field E_i and the electric displacement D_i. The stress and strain are second order tensors, while the electric field and the electric displacement are first order. The superscripts T, D; E signify that these quantities are measured at zero stress, or zero current, or constant electric field, respectively.

Types of Piezo-Electric Materials

Piezocrystals and Piezoceramics – For a long period, from the discovery of piezoelectricity in 1879 and until W.W.II, the only piezo-electric materials widely used were quartz crystals (SiO_2) and Rochelle salt. The latter is a natural ferroelectric material, that posses an orientable domain structure that aligns under an external electric field and thus enhances its piezoelectric response. The discovery of ferroelectric piezoceramics started in the late 1940's with barium titanate ($BaTiO_3$). In 1954, lead zirconate titanate ($PbZrO_3$) with its even stronger piezoelectric effects, started to be produced. Under the commercial name of PZT, this material is one of the most widely used piezoceramics. To date, many PZT formulations exist, the main differentiation being between "soft" and "hard" PZT varieties. Other currently available piezoceramics are lead titanate, lithium niobate; lithium tantalate.

Piezopolymers – Polyvinylidene fluoride (PVDF), also abbreviated PVF_2, is a polymer with strong piezoelectric and pyroelectric properties. In the α phase, PVDF is not polarized and is used as a common electrical insulator among many other applications. To impart the piezoelectric properties, the α phase is converted to the β phase and polarized. Stretching α-phase material produces the β phase and metallization of the surface; the application of a strong electric field provides the permanent polarization for the piezoelectric properties. The flexibility of the PVDF overcomes some of the drawbacks associated with the brittle piezoelectric ceramics. As a sensor, it provides higher voltage/electric field in response to a mechanical stress. Its easy formability, along with this property, makes it superior to ceramics as a sensor. The g-constants, which represent the voltage generated in response to a mechanical stress, are typically 10 to 20 times that of piezoceramics. PVDF film also produces an electric voltage in response to infrared light; the pyroelectric coefficient defines this relationship. The constitutive relations for PVDF can be described as

$$\begin{aligned} S_{ij} &= s^E_{ijkl} T_{kl} + d_{kij} E_k + \alpha^E_i \theta \\ D_j &= d_{jkl} T_{kl} + \varepsilon^T_{jk} E_k + \widetilde{D}_m \theta, \end{aligned} \tag{3}$$

where, apart from the notations already described, θ is the absolute temperature, α^E_i is the coefficient of thermal expansion under constant electric field; $\widetilde{D}_j$ is a the electric displacement / temperature coefficient.

Applications of Piezoelectric Materials to Stress Sensing

Piezoelectric materials produce a voltage (open circuit operation), or an electric displacement (close circuit operation) when subjected to mechanical strain. In stress and pressure sensing, the PVDF materials have been found to give a better electrical response then piezoceramics due to the inherent low modulus that results in a large strain under small applied stress. A PVDF sensor is used like a strain gage but does not require a conditioning power supply. The output signal is also comparable to that of an amplified strain gage signal. This high sensitivity is due to the low thickness of the typical PVDF film (1 mil). Because of its good sensor properties (i.e., the high g constant), light weight, flexibility; toughness, PVDF is used in numerous sensor applications. PVDF is very commonly used in switching applications, where it develops a voltage in response to a pressure (it develops approximately 10 V with finger pressure). Keyboards make use of this property. One surface of the film is completely metallized, whereas the other surface is metallized only in areas corresponding to the keys and their associated signal-detection wires. The film is stretched under the metallic keyboard with recesses at various key locations. The finger pressure bends the film, producing a signal corresponding to the specific key, which is then used to perform appropriate tasks.

PVDF film is also used as for acoustic applications both as sensors and as actuators. It is used in compact microphones, earphones; loudspeakers. Specially shaped PVDF sensors have also been used as modal sensors for sensing the natural modes of vibration of structural elements such as beams, plates and cylinders.

Many surveillance systems make use of the pyroelectric properties of the PVDF sensor. This application typically uses two closely coupled sensors. Much like the ambient-temperature compensation scheme used with strain gauge measurements, only one sensor is exposed to the infrared emissions, whereas both sensors respond equally to other stimuli, such as noise, vibration; global temperature changes. Signals from both sensors are combined to detect the infrared signal alone.

Applications to Health Monitoring – The electro-mechanical impedance technique utilizes the direct and the converse electro-mechanical properties of piezoelectric (PZT) materials, allowing for the simultaneous actuation and sensing of the structural response. The variation of the electro-mechanical impedance of a PZT sensor-actuator interrogator intimately bonded to the structure is monitored over a large frequency spectrum in the high kHz frequency band. Figure 11 presents the experimental set-up and the conceptual system diagram for health monitoring a structure using the electro-mechanical impedance approach.

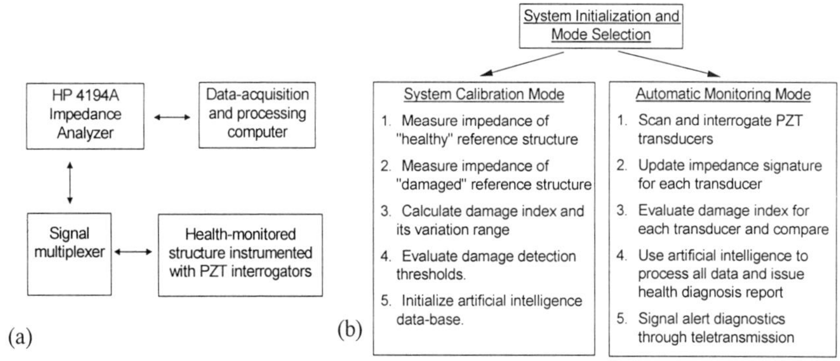

Figure 11 – Principles of the electro-mechanical impedance technique: (a) Equipment schematic; (b) conceptual diagram of the automated monitoring system.

Two basic ingredients are essential of this method: an array of piezo-electric PZT sensor-actuators applied to the monitored structure; a high-precision impedance analyzer coupled to a data-acquisition computer. The size of the PZT sensor-actuators is typically small (less than 0.5 sq. in., 0.01 in. thick), allowing for non-intrusive installation in the monitored structure. The PZT sensor-actuator is excited electrically by high-frequency low-power voltage; its complex electro-mechanical impedance is measured over a given frequency range. Figure 12a shows the frequency response diagrams obtained during a typical laboratory demonstration of the electromechanical impedance technique. Figure 12b presents a typical damage index diagram that distinguishes between damaged and undamaged locations during the health-monitoring process.

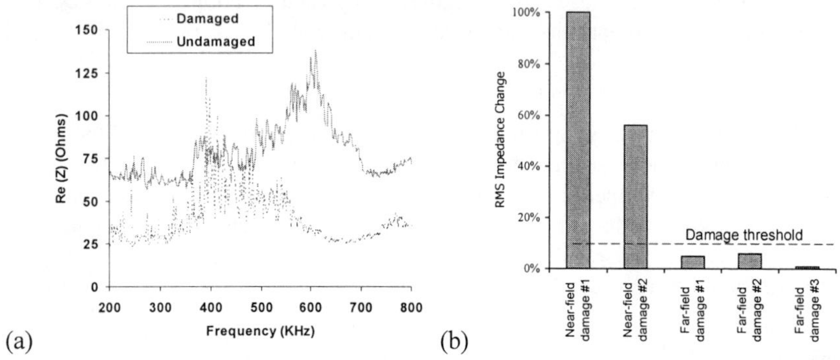

Figure 12 – Typical damage detection results: (a) impedance signature pattern; (b) damage index comparison (Giurgiutiu and Rogers, 1997a).

The high resolution of this incipient damage detection technique is ensured through the intimate electro-mechanical coupling between the electrical impedance response of the piezo-electric sensor-actuator affixed to the structure and the local mechanical impedance of the adjacent material present in the structure and in the structural joints. Localization of the sensing area ensures sensitivity of the impedance signature only to damage and/or structural changes in the near-field area of the PZT sensor-actuator. The insensitivity of the methods to far-field changes ensures good rejection of the unwanted far-field information and prevents the method from giving "false alarms" due to changes that are part of the normal structural usage (boundary conditions, mass changes, service loads, etc.)

The E/M impedance technique is still in its infancy. However, its potential for application is very large. Proof-of-concept demonstrations for civil engineering applications include the use in the laboratory monitoring of a model steel bridge joint structure (Ayres, Chaudhry; Rogers, 1996); of composite overlays masonry repairs (Quattrone and Berman, 1997); in a full-scale Navy pier composite upgrade (Quattrone and Berman, 1997).

Advantages and Limitations of Piezoelectric Materials – Piezoelectric ceramics, e.g., PZT, have a large piezoelectric response, but also possess certain limitations. The most obvious limitations are associated with non-linearity, hysteresis, creep, depoling, electrical breakdown; Curie temperature.

Hysteresis is the phenomenon that occurs when, upon removal of the electric field, a certain mechanical strain remains in the piezoelectric material. The cause of the hysteresis is the energy dissipation due to internal sliding events in the polycrystalline piezoelectric body. The degree of hysteresis depends on the field level, the cycle time; the material of the piezoelectric actuator. Hysteresis is often measured as a percentage of the total strain; typically in the range 0.1 to 10% (Figure 13). Since the energy dissipated in the hysteresis cycle appears in the form of heat, it is a concern in high frequency operations, where it can lead to excessive temperatures.

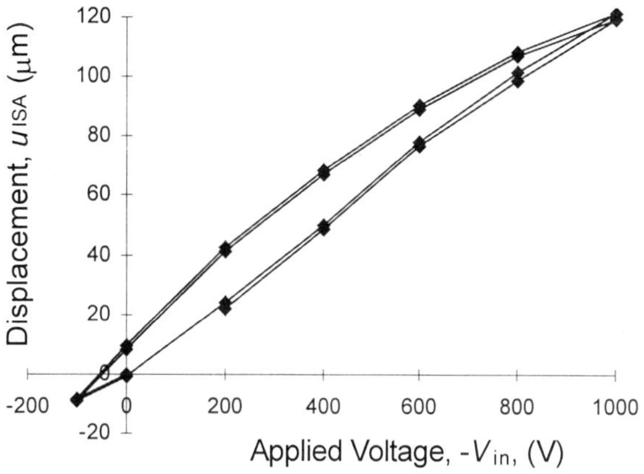

Figure 13 The induced-strain linear displacement vs. applied voltage curve for PI P-245.70 PZT translator (Giurgiutiu and Rogers, 1997).

Creep refers to plastic deformations that take place (under constant loads) over time. In piezoelectric materials, it is the prolonged application of electric field that causes the induced mechanical strain to grow with time. Creep is a concern when a piezoelectric sensor is being subjected to a constant force for a long time, or when a piezoelectric actuator is extensively subjected to high dc voltages. Creep is not a concern under ac operation.

Depoling refers to the situation in which, when a high electric field is applied to the piezoelectric opposite to the poling direction, its piezoelectric properties may degrade or it may get poled in the opposite direction. The field that causes depoling is called a coercive field. It is important to note that a high stress level can also cause depoling. For ac operation, the coercive field limit is higher because the duration of the applied field is shorter. The coercive field limit is generally available from the manufacturer; if not, it can easily be determined through a simple experiment.

Electrical breakdown happens when a strong electric field applied in the poling direction before produces loss of the electrical insulation properties of the material allowing flow of electricity through the body of the piezoelectric or across its edges. Electrical breakdown destroys the piezoelectric properties. The electrical breakdown field is quite high and can typically be twice as much as the coercive field. Arcing can also occur at the edges of a piezoelectric wafer at low field levels, but it should not be confused with electrical breakdown. This arcing is generally due to the debris left adhering to the edges after cutting. The arcing vaporizes the debris, thereby acting as a self-cleaning action.

Curie temperature is the maximum operating temperature beyond which the material loses all its piezoelectric properties. The operating temperature should generally be much lower than the Curie temperature because at temperatures close to the Curie temperature, depoling is facilitated, aging and creep are accelerated; the maximum safe mechanical stress also decreases. For typical PZT material, this temperature is about 350° C.

Smart Composites by Tagging

Passive and active tagging are sensing techniques that involve adding "tagging" particles to materials. Embedded piezoelectric, magnetostrictive, electrostrictive, or ferromagnetic particles can be used to provide inherent information about the in-process or in-service state of adhesives and polymers. Tagging offers the advantage of distributed in-situ sensing, which is not possible with many types of sensors. If the material system is properly designed, the tagging particles will not adversely affect the properties of the host material, as will other types of embedded sensors. Passive techniques involve sensing the distribution of the particles. An example of a passive technique is adding magnetic particles to an adhesive and using an eddy current probe to detect voids in a tagged adhesive bond. Active tagging involves exciting the sensing particles and measuring the response of the host material. Applying an alternating magnetic field to a polymer tagged with magnetic particles and measuring the resulting force would be an example of active tagging. Applications of passive and active tagging techniques include characterization of adhesive bonds, cure monitoring, intelligent processing, nondestructive materials evaluation, damage detection; in-service health monitoring.

A sustained effort to develop tagged composites for in-situ and in-service health monitoring and damage detection is taking place under a CRADA agreement between the Construction Engineering Research Laboratories of the US Army Corps of Engineers and the Composites Institute of the Society of Plastics Industries with the cooperation of several universities.

Tagging Principles with Electric and/or Magnetic Fields – Many of the established NDE techniques for crack detection and damage assessment rely on electric and/or magnetic principles. For example, the Magnaflux method applies a magnetic field through a ferromagnetic part of machinery and relies on the flux leakage around an open crack to collect fine powders of magnetic particles and produce a visual indication of the crack location. Another well established NDE method for crack detection relies on the eddy currents induced by a probe traveling above the metallic piece under inspection to create a reflected magnetic field whose phase and amplitude modifies when a crack is present. However, most polymeric matrix composites are not ferromagnetic. Hence, ferromagnetic NDE methods cannot be directly applied.

Additionally, with the exception of carbon fiber composites, polymeric composites are not electrically conductive; eddy currents cannot be directly induced in them. These limitations can be overcome by incorporating in the composites trace quantities of tagging materials that will impart magnetic susceptibility and/or electric conductivity. Of these two alternatives, the latter is less attractive since one of the well praised qualities of the polymeric composites is their insulating nature. Tagging with ferromagnetic particles has been extensively studied (Giurgiutiu, Chen, Lalande, Rogers, Quattrone; Berman, 1996). In a first stage of the research, sample specimens incorporating up to 5% by weight of ferromagnetic tags (magnetite, nickel-zinc ferrite, manganese-zinc ferrite, Lignosite, etc.) were produced by a number of industrial partners. The composite used in these experiments was of fiberglass type, with a selection of resins (polyester, epoxy, vinyl ester, etc.) used in the matrix. These specimens were subjected to two types of NDE processes: (a) eddy current interrogation; (b) electromagnetic excitation of structural vibrations. The eddy currents method was not found very effective in detecting internal defects (delamination) since the presence of ferromagnetic tags did not ensure a path for the eddy currents inside the composite. However, this method was found to be effective in detecting defects in carbon fiber composites which are conductive and thus create a path for the eddy currents to form. The electromagnetic excitation of vibrational response was found effective in producing marked difference in the frequency response spectrum due to the presence of simulated defects (Figure 14).

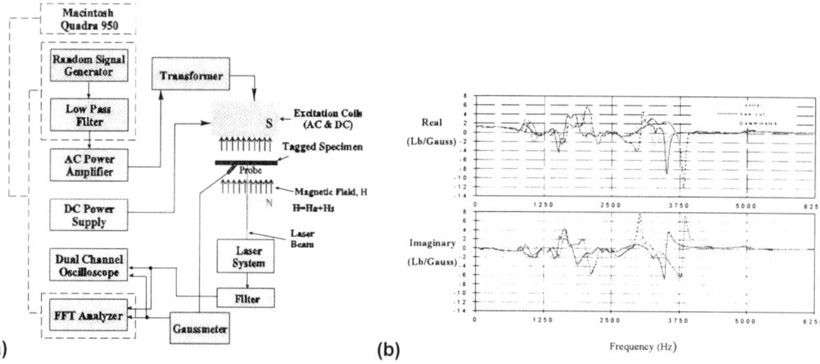

Figure 14 – Active tagging interrogation of ferromagnetically tagged smart composites specimens: (a) schematic of the experimental set-up; (b) frequency response diagrams for control and damaged specimens

Further investigations were performed on full-scale composite specimens. Pultruded C-shaped composite ladder rails were produced with ferromagnetic tagging particles incorporated in the resin material The tagged ladder rails were subjected to contactless electromagnetic excitation. The resulting vibrational response was detected with a laser Doppler velocimeter aimed through a small port in the electromagnet yoke (Figure 15a). Thus, the detection method was, again, contactless. A typical vibrational frequency response, investigate in the kHz range, is shown in Figure 15b. This figure also shows the frequency response curve of a specimen with simulated defects in the form of a central web delamination. The marked difference between the two spectra can be processed to yield significant information about the structural health and the presence of structural defects.

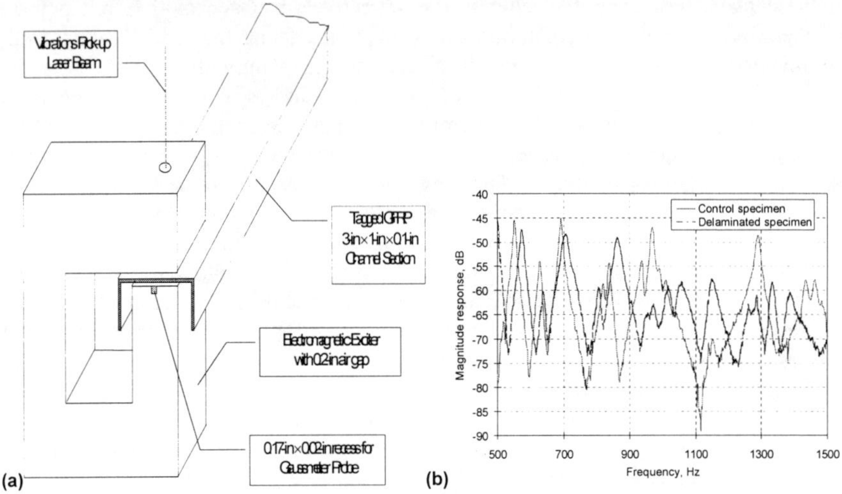

Figure 15 – Active tagging interrogation of full-scale ferromagnetically tagged smart composite C-rail elements: (a) experimental set-up; (b) frequency response diagrams for control and damaged specimens

Two tagging types were used during these experiments: (a) magnetite and (b) Lignosite. The magnetite tagging gave a better response but decreased somehow the strength properties of the composite, probably due to the roughness and hardness of these particles. The Lignosite tagging blended much better into the resin-fiber material system of the composite, but its overall magnetic response was not as good as that of the magnetite.

Tagging Principles with UV and Fluorescent Light – Substances that respond to ultraviolet (UV), fluorescent or infrared (IR) light are commonly used in tagging explosives and other high-security materials. The underlying principle of these tags is that luminescence is created when the material is bombarded with high-energy quanta of electromagnetic waves. The same principle can be applied to structural composites to sense the presence of damage and delaminations.

Magnetostrictive Particle Tagging – Magnetostrictive materials present the quasi-piezomagnetic effect, i.e. they create a magnetic field when subjected to mechanical strain. When magnetostrictive particles (e.g., TERFENOL powders) are embedded in a polymeric matrix, they can become in-situ strain and stress sensors. If the tagged matrix is subjected to mechanical stress, a state of strain is realized in the heterogeneous medium made up of the matrix and the TERFENOL tags. Hence, the TERFENOL tags generate a magnetic field that can be measured with a Gaussmeter. If the tagged matrix material is used in a fiber/resin composite system, the recorded magnetic response will reflect the state of stress in the composite. The method could be especially effective in detecting cracks, since the stress and strain at the crack tip can be many times larger that the background stress and strain in the composite material. Thus, a simple reading of the magnetic field could detect regions with increased magnetic flux values that are indicative of the presence of internal cracks.

Work in this area has been performed by Rogers et al (1995), by White and Albers (1996) and White, Albers and Quattrone (1996). Rogers et al showed that the presence of a crack in a TERFENOL-tagged polymeric resin subjected to mechanical stress can be detected with a Gaussmeter. Extensive work on tagging composite materials with magnetostrictive powders has been accomplished by White, Albers and Quattrone (1996). Experiments with TERFENOL powder embedded in polymers and polymer composites were performed.

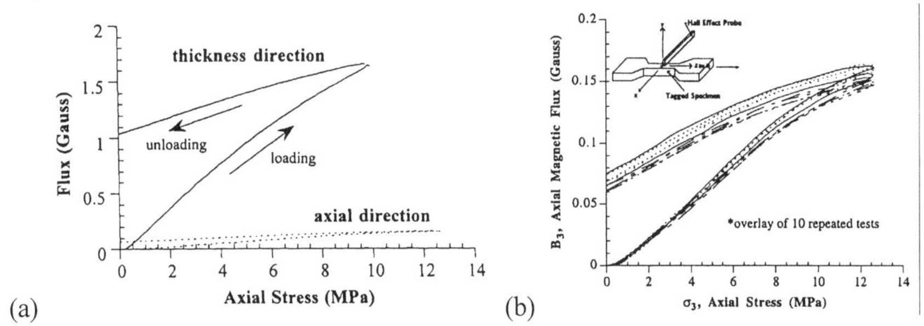

Figure 16 – Correlation between applied stress and magnetic field in magnetostrictively tagged resin: (a) comparison of measurements made in the thickness and axial directions (White, 1999); (b) repeated tests showing consistent response (Quattrone, Berman, and White, 1998)

The magnitude of stress in TERFENOL-tagged materials could be correlated with the strength of the magnetic field detected on the surface of the material (Figure 16). Thus, areas of high stress or concentrated stress are reflected in the magnetic signature of TERFENOL-tagged materials (Travillion *et al.,* 1999). Experiments with magnetostrictive tagging of glass-fiber reinforced composite beams were demonstrated by Giurgiutiu *et* al. (1999). These experiments extended the correlation between stress and magnetostrictively induced magnetic field to bending loading of full-scale composite members. Further work on this method is under progress towards field portable NDE equipment.

Advantages and Limitations – The main advantage of smart tagged composites lays in the fact that this approach will give composites properties that were otherwise unattainable. For example, a polymeric composite is not magnetic, but a tagged composite can be made to respond to magnetic fields. These added properties extend the possible options for in-situ structural health monitoring and damage detection of this class of materials. One of the main concerns about tagged composites is related to the long term durability of the complete system. Long term durability studies are still to be performed and performance indicators have still to be determined in comparison to more conventional material systems. The other concern is the initial and life-cycle cost of these products. Both installation costs and in-service costs for implementing the monitoring functions need to be considered and assessed. Quattrone and Berman (1998) also see as an important question to be addressed the need for creating the appropriate structure to coordinate the information passage between users, developers and installers of smart composites.

Electro/Magnetostrictive Materials for Vibration Control

Principle of electro/magnetostriction – Electro- and magnetostriction is the property of changing the linear dimension under the effect of an externally applied electric or magnetic field. The electro and magnetostriction effects are quadratic, i.e. they depend on the square of the applied field. For this reason, electro and magnetostriction, in its rough form, is not a bi-directional phenomenon. Under the effect of an external field, an electro/magnetostrictive material will generally expand in the direction of the field lines, irrespective of their sense. Electro/magnetostrictive materials can be made to behave quasi-linearly by applying a bias field. This shifts the operating point such that a non-zero expansion is obtained under zero external signals. A subsequent external signal, which adds (positive) or subtracts (negative) from the bias field, will result in a positive or negative displacement with respect to the bias position.

Electrostrictive Materials – A ferroelectric ceramic with interesting properties is the compound lead-magnesium-niobate, or PMN. This material exhibits a strong electrostrictive effect. Electrostriction involves a nonlinear electromechanical coupling for which the material develops a strain proportional to the square of the polarization. The strain versus applied field curves for electrostrictive material is shown in Figure 17. Electrostrictors exhibit considerable less hysteresis; the phenomenon of depoling does not exist. However, commercially available electrostrictive actuators as internally biased such that reverse polarity of the excitation field is unacceptable.

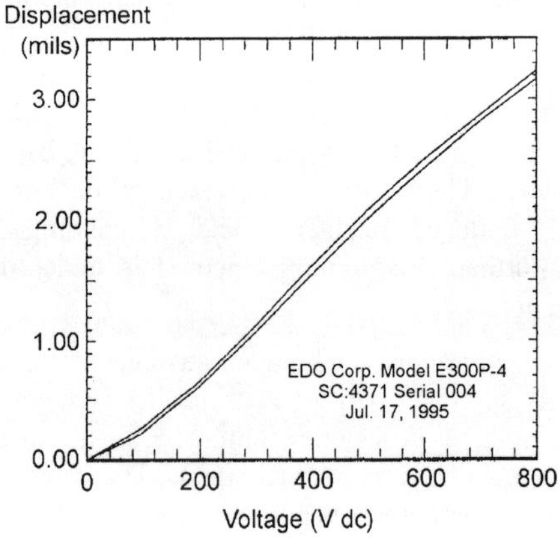

Figure 17 – Voltage-displacement curve for the EDO Corporation model E300P PMN device (1 mil = 25.4 μm).

For PMN-PT (lead magnesium niobate doped with lead titanate), induced strains of approximately 1200 μstrain have been reported. The nonlinear electromechanical coupling of this material can be used to develop transducers with tunable sensitivities. Because the slope of the repeatable electric field-to-strain relationship varies, the transduction sensitivity, whether the ceramic is used as an actuator or a sensor, can be controlled by adjusting the dc-bias electric field about which the electrostrictive transducer operates. The disadvantage of the nonlinear (quadratic) response is the reduced actuator authority for ac operations. Furthermore, the electromechanical response is frequency dependent and is restricted to a rather small temperature range in which the dielectric permittivity is large. The deformation for low applied electric fields is small.

The constitutive relations for electrostrictive actuators are written similarly to those for piezoelectric actuators but second-order terms are included, e.g.,

$$S_{ij} = s_{ijkl}^{E} T_{kl} + d_{kij} E_k + M_{klij} E_k E_l \,. \tag{4}$$

In this equation, the first two terms are the same as those used to describe the piezoelectric constitutive law: Hook's law and the converse piezoelectric effect. The third term represents the electrostriction effect. The components of M_{klij} are the electrostrictive coefficients. Equation (4) indicates that electrostriction appears as quadratic addition to the linear piezoelectric effect. In fact, the two effects are separable because the piezoelectric effect is possible only in noncentrosymmetric materials, whereas the electrostrictive effects are not limited by symmetry and are present in all materials. In addition to the direct electrostrictive effect, the converse electrostrictive effect also exists.

Magnetostrictive Materials – Magnetostrictive materials expand in the presence of a magnetic field, as their magnetic domains align with the field. The magnetostrictive effect was first discovered in nickel in 1840 and was later found to be present in cobalt, iron; their alloys. However, magnetostriction in these materials was relatively small, on the order of 50 μstrain. Strains on the order of 10^{-2} were observed in the rare-earth elements terbium (Tb) and dysprosium (Dy) at very low temperatures (below 180K). Scientists working with the Ames Laboratory and the Naval Ordnance Laboratory (now the Naval Surface Weapons Center) developed Tb-based compounds, such as the commercially available Terfenol-D ($Tb_{.3}Dy_{.7}Fe_{1.9}$), which have magnetostriction of up to 2000 μstrain at temperatures up to 80° C and higher.

Magnetostriction is dependent on applied stress. When there is no applied compressive stress, the "jump" in magnetostriction does not occur while, above a certain compressive stress value, magnetostriction decreases. The reason for this unusual behavior is that the applied stress changes the alignment of the magnetic domains. The larger the rotation that occurs, the larger the observed strain. With sufficient compression stress (around 7.6 MPa) and no applied magnetic field, a large percentage of the magnetic domains are oriented perpendicular to the direction of applied stress. When magnetic field is applied, the domains rotate to align with the applied field which is in the direction of the applied stress. Note that as the pre-stress is further increased (e.g., 18.9 MPa), large magnetic fields are required to produce the same strains, since the material performs more mechanical work when pushing against a larger stress.

A simple two-dimensional model to explain magnetostriction in TERFENOL was developed by Clark (1992). In the first stage of magnetostriction, the domains are oriented perpendicular to the applied compressive stress at an initial magnetic field of zero. Note that the material consists of alternating parent and twinned phases. Magnetostriction does not occur until a critical magnetic field is reached, when the parent phase "jumps" to a new direction and large strains are observed. Magnetostriction continues to increase as the applied field is increased beyond this critical value, as both the parent and twin rotate toward the direction of the applied field. This model accurately predicts the observed magnetostriction values.

The following actuator and sensor equations describe the constitutive behavior of TERFENOL material:

$$\begin{aligned} S_{ij} &= s^E_{ijkl} T_{kl} + d_{kij} H_k + M_{klij} H_k H_l \\ D_j &= d_{jkl} T_{kl} + \mu^T_{jk} H_k \end{aligned} \tag{5}$$

where, in addition to the already defined variables, H_k is the magnetic field intensity; μ^T_{jk} is the magnetic permeability under constant stress. Of course, the coefficients d_{kij} and M_{klij} are defined in terms of magnetic units. The magnetic field intensity H in a rod of length L is related to the current in the surrounding coil (with n turns per unit length) through the following equation:

$$H = nI \tag{6}$$

Design of damping systems with these materials – Piezoelectric, electrostrictive and magnetostrictive materials can be used to produce structural damping. Two mechanisms are available: (a) dissipate energy directly through the active material electromechanical interaction; (b) use the active material to enhance the damping properties of a conventional damping material.

The energy can be dissipated using the electromechanical interaction of the active materials in the following way. The active material is connected to the structure undergoing vibration. The deformation of the active material is made to follow the deformation of the structure. As the structure deforms, the active material takes up the strain and transforms it into an electrical or magnetic field, as appropriate for the active material used. If the material is electroactive (piezoelectric or electrostrictive), the induced electric field is used to drive current into an external resistance thus dissipating the energy through Ohmic heating. If the material is magnetoactive (e.g., magnetostrictive), the resulting magnetic field is transformed into induced current into a coil surrounding the magnetostrictive material. The induced current is passed through a ballast resistance that dissipates the energy through Ohmic heating.

The active material can be used to enhance the damping properties of a conventional damping material through the "constraint layer damping effect". Conventional vibration damping treatments utilize the dissipation properties of viscoelastic materials that are applied to the structure. As the structure vibrates, the viscoelastic material is cyclically deformed and dissipates energy. The energy dissipation in viscoelastic materials is mainly a shear strain effect. This effect can be enhanced if on top of the damping layer one adds a layer of active material. The active material is made to deform in opposite phase with respect to the base structure. Thus, the damping layer in between the active material and the base structure would be subjected to much larger differential strain than if no active constraint layer were present.

Advantages and limitations – The main advantage of electro/magnetostrictive materials over the conventional piezoelectric/magnetic materials is their very low hysteresis. This could be especially beneficial in dynamic applications. The main disadvantage of electro/magnetostrictive materials is their nonlinearity. However, by applying a bias field, quasi-linear behavior around the bias operating point can be obtained.

Actuating Materials

The electro- and magnetoactive materials previously described for their sensory applications can also be used for actuating applications. For example, a piezoceramic material can be used as a strain sensor by taking advantage of the direct piezoelectric effect that converts a mechanical action into an electric charge. But the same material can be also used for actuation applications by using the converse piezoelectric effect that converts an electrical field into a mechanical strain. Such solid-state actuators that utilize the strain induced in the active material through electric or magnetic actions to produce an output displacement are called induced-strain actuators.

Piezoelectric and Electrostrictive Materials

The basic properties and constitutive relations of piezoelectric and electrostrictive materials have been already discussed in detail. Here, attention is focused on the use of these materials for the construction of induced-strain actuators.

The use of piezo-materials as actuators – An electroactive solid-state actuator consists of a stack of many layers of electroactive material (PZT or PMN) alternatively connected to the positive and negative terminals of a high voltage source (Figure 18). Such a PZT or PMN stack behaves like an electrical capacitor. When activated, the electroactive material expands and produces output displacement. Typical strains for electroactive materials are in the range 750-1200 μm/m.

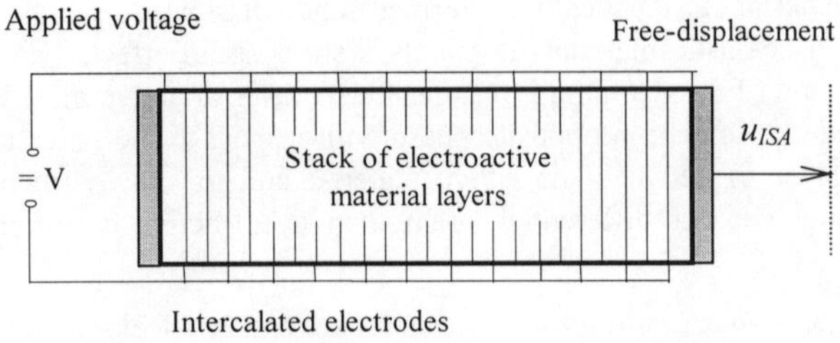

Figure 18 Induced strain actuator using a PZT or PMN electroactive stack

The PZT or PMN stacks are constructed by two methods. In the first method, the layers of active material and the electrodes are mechanically assembled and glued together using a structural adhesive. The adhesive modulus (typically, 4-5 GPa) is at least an order of magnitude lower than the modulus of the ceramic (typically, 70-90 GPa). This aspect may lead to a stack stiffness that is significantly lower than the stiffness of the basic ceramic material. In the second method, the ceramic layers and the electrodes are assembled in the "green" state; then fired together (co-fired) under a high isostatic pressure (HIP process) in the processing oven. This process ensures a much stiffer final product and, hence, a better actuator performance. However, the processing limitations, such as oven and press size, etc., limit the applicability of this process to small size stacks only.

The PZT and PMN stacks are surrounded by a protective polymeric or elastomeric wrapping. Lead wires protrude from the wrapping for electrical connection. Steel washers, one at each end, are also provided for distributing the load into the brittle ceramic material. When mounted in the application structure, these stacks must be handled with specialized knowledge. Protection from accidental impact damage must be provided. Adequate structural support and alignment are needed. Mechanical connection to the application structure must be such that tension stresses are not induced in the stack since the active ceramic material has very low tension strength. Hence, the load applied to the stack must always be compressive and perfectly centered. If tension loading is also expected, adequate pre-stressing must be provided through springs or other means.

Design of effective piezo actuators – The design of effective piezo actuators is particularly difficult due to the small displacement generated by these materials. Typical maximum strains are in the range of 0.1% which, for a usual length of 100 mm (≈ 4") becomes a no-load displacement of 0.1 mm (0.001"). However, the forces that can be generated by a piezo actuator can be very large; are only limited by the inherent stiffness and compressive strength of the piezo material; by its effective area. For most practical applications, displacement amplification of the induced strain displacement is employed. The effective design of the displacement amplifier can "make or break" the practical effectiveness of a piezo actuator. Giurgiutiu and Rogers (1996a, 1996b) performed extensive studies of the effective design of displacement amplified induced-strain actuators and showed that the most important parameter that needs to be optimized during such a design is the energy extraction coefficient defined as the ratio between the effective mechanical energy delivered by the actuator and the maximum possible energy that can be delivered by the actuator in the stiffness-match condition. (The stiffness match condition is the situation in which the internal stiffness of the actuator equals the external stiffness of the application).

Of extreme significance in the design of an effective induced-strain actuator is the amount of energy available for performing external mechanical work. Comparison of the energy density capability of a selection of induced strain actuators commercially available on the open market was performed by Giurgiutiu, Chaudhry and Rogers (1996a) for the static regime and by Giurgiutiu and Rogers (1996b) for the dynamic regime. Absolute energy values and the energy density per unit mass and unit volume were computed and compared. Figure 19 presents such comparison charts for the energy density per unit volume for static (Figure 19a) and dynamic (Figure 19b) operation.

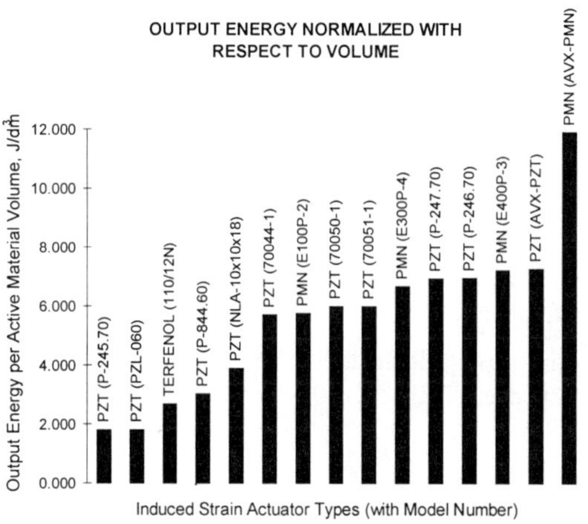

(a)

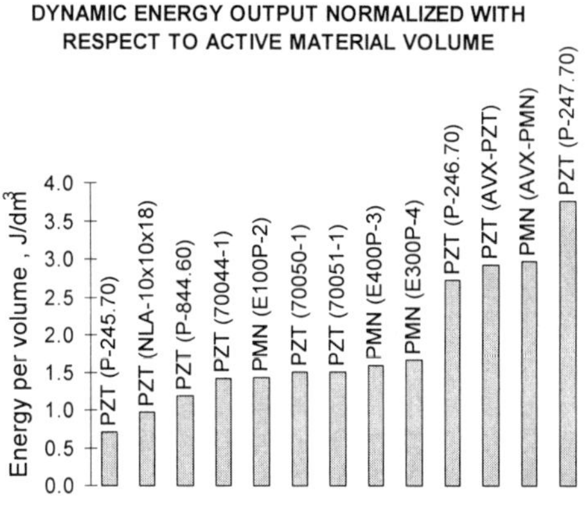

(b)

Figure 19 – Specific energy output capabilities of a selection of commercially available induced-strain actuators: (a) static operation; (b) dynamic operation.

Applications – PZT and PMN actuators can be used as linear actuators in a stack configuration (this utilizes the electrostrictive effect parallel to the direction of the applied field) or to induce bending in the bonded structure (this utilizes mechanical strain induced transverse to the direction of the applied field). Bonded electrostrictive actuators have been used successfully to control the shape of deformable mirrors. In the stack configuration, they have been used for impact dot-matrix printing. Advantages over conventional electromagnetic actuators included order-of-magnitude higher printing speeds, order-or-magnitude lower energy consumption; reduced noise emissions. A tunable ultrasonic medical probe composed of electrostrictors embedded in a polymer has also been developed.

Advantages and limitations – This simple analysis shows that very large output power densities can be extracted from induced-strain devices. The practical limitations on the power delivery capabilities of piezo-active materials stem from three directions:

1. The capability of the power supply to deliver the required reactive input power which can be quite large (kVA range).
2. The dissipation of the heat generated in the active material due to internal losses. These losses can be in the range 5%-10% of the nominal reactive input power.
3. The mechanical resonance of the piezo driven system that sets an upper limit on the frequency variation.

Many of the electric-supply difficulties listed above can be avoided by using piezo-magnetic devices, such as TERFENOL-D based induced strain actuators.

Magnetostrictive Materials

The basic properties and constitutive relations of magnetostrictive materials have already been discussed in detail. Here, the focus will be on the use of these materials for the construction of induced-strain actuators.

Design of magnetostrictive actuators – Magnetostrictive materials can be used to produce an effective actuator. Figure 20 shows the typical layout of a magnetostrictive actuator. A magnetoactive solid-state actuator consists of a TERFENOL bar inside an electric coil and enclosed into an annular magnetic armature (Figure 20). End caps ensure that the magnetic circuit is closed. In this arrangement, the magnetic field is strongest in the cylindrical inner region filled by the TERFENOL bar. When the coil is activated, the TERFENOL expands and produces output displacement. The TERFENOL-D bar, the coil; the magnetic armature, assembled between two steel-washers and put inside a protective wrapping, form the basic magnetoactive induced-strain actuator. Prestressing end springs may also be present.

The TERFENOL material has been shown to be capable of strains up to 2000 μm/m, but with highly nonlinear behavior. Practical strains employed by the manufacturers of TERFENOL actuators are in the quasi-linear behavior range of 750-1000 μm/m. A typical large-power magnetostrictive actuator (ETREMA AA-140J025) is 200 mm long, weighs ≈1 kg and can produce 0.140 mm output displacement.

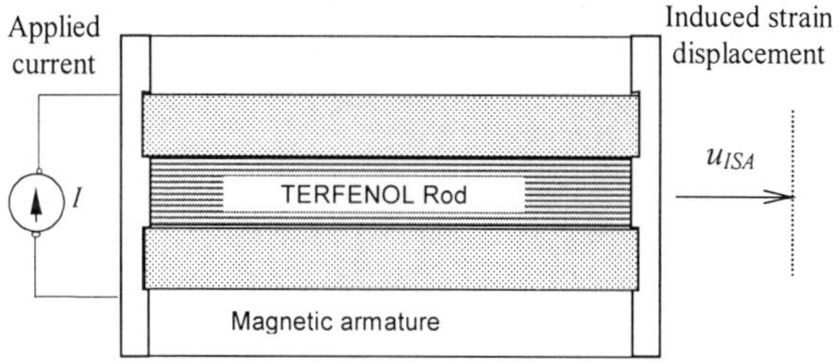

Figure 20 – Schematic of a magnetostrictive (TERFENOL) solid-state actuator

Advantages and limitations – The main advantage of magnetoactive induced-strain actuators over their electroactive counterparts resides in the power supply requirements. The magnetoactive actuator is a current-driven device; hence can operate at voltages much lower than the electroactive actuators. Since, at present, the low-voltage power supply technology is more developed and cheaper than the high-voltage technology, the use of magnetostrictive actuator is easier and more attractive.

The main limitation of the magnetoactive induced-strain actuators stems from the necessity for additional elements in the actuator construction. While a bare-bones electroactive actuator need not contain anything more than just the active material stack, the bare-bones magnetoactive actuator needs always to include the energizing coil and the magnetic armature. For this fundamental reason, the power density (either per unit volume or per unit mass) of magnetoactive induced-strain actuators will always remain below that of their electroactive counterparts.

Shape Memory Alloys

The first observation of the shape-memory effect (SME) was made in 1932 with gold-cadmium. The phase transformation associated with the shape-memory effect was later discovered in 1938 with brass. It was not until 1962 that Buehler *et al.* at the Naval Ordnance Laboratory (NOL) discovered a series of nickel-titanium alloys that demonstrated this shape-memory effect. The shape-memory alloy discovered by Buehler et al., later named Nitinol, has been commercially available ever since. Many materials are known to exhibit the shape-memory effect. They include the copper alloy systems of Cu-Zn, Cu-Zn-Al, Cu-Zn-Ga, Cu-Zn-Sn, Cu-Zn-Si, Cu-Al-Ni, Cu-Au-Zn; Cu-Sn, the alloys of Au-Cd, Ni-Al; Fe-Pt; others. The most common of the shape-memory alloys or transformation metals is the Nickel-Titanium alloy, Nitinol.

Physical basis of shape memory and the different types of shape memory materials – The shape-memory effect (SME) can be described as follows: an object in the low-temperature martensitic condition, when plastically deformed under no external stresses, will regain its original (memory) shape when heated. This process is the result of a reverse martensitic transformation that takes place during heating. The martensitic transformation, which is the essence of the response of shape-memory alloys, may be illustrated simply by the change in martensite volume fraction with respect to temperature. The four important transition temperatures are martensite-finish (M_f), martensite-start (M_s), austenite-start (A_s); austenite-finish (A_f).

Nickel-titanium alloys of proper composition exhibit unique mechanical "memory" or restoration force characteristics. The shape-recovery performance of Nitinol is phenomenal. The material can be plastically deformed in its low-temperature martensite phase and then restored to the original configuration or shape by heating it above the characteristic transition temperature (the characteristic temperature can be varied from -50 to 166°C). This unusual behavior is limited to NiTi alloys having near-equiatomic composition. Plastic strains of typically 6 to 8% may be recovered completely be resistively heating the material so as to transform it to its austenite phase. Restraining the material from regaining its memory shape can yield stresses of 700 MPa.

In addition to the special characteristics noted above, shape memory alloys have some other special properties. The change in the Young's modulus of shape-memory alloys is very different from that of conventional metal materials. For most metals, the Young's modules decreases as the temperature increases. However, the Young's modules of shape-memory alloys increases within the phase transformation temperature range for the heating process. The Young's modulus of Nitinol increases three to four times from temperatures below M_f to temperatures above A_f. Another property of shape-memory alloys that has attracted a great deal of attention is the damping characteristic (i.e., the internal friction characteristics of shape-memory alloys). The damping properties of shape-memory alloys may be exploited for passive and adaptive dynamic control applications. Shape-memory alloys are different from other materials in other aspects too. The influence of the phase transformation on the electrical resistance is one example. An evolution of the internal resistivity of Nitinol with its phase transition and a clear distinction between the heating and cooling process has been observed. Measuring the electrical resistance of shape-memory alloys is often used to determine the phase transition temperatures.

Constitutive Relations of Shape-Memory Alloys – The mechanical behavior of shape-memory alloys is closely related to the microscopic martensitic phase transformation; the constitutive relations developed for ordinary materials such as Hook's law and plastic flow theory are not directly applicable to shape-memory alloys. Therefore, new constitutive relations, which take into consideration the phase transformation behavior of SMA, have been developed. Several constitutive relations have been developed for shape-memory alloys over the last 20 years (Cory, 1978; Muller, 1979; Tanaka and Nagaki, 1982). While each is unique, none of the constitutive relations has a strong experimental justification; each possesses distinct limitations (Liang, 1990). There are two approaches to developing material constitutive relations. One is the macroscopic phenomenological method that requires a significant amount of experimental work; the other is the microscopic or physical method that derives material constitutive relations based on fundamental physical concepts. Combining both approaches takes advantage of each method and yields a more accurate constitutive model capable of predicting and describing the material behaviors of SMA.

Modeling of SMA with Internal Variables – Tanaka and his collaborators (1982, 1985, 1988) developed a model based on the concept of the free-energy driving force. Tanaka's model considers a one-dimensional metallic material of length L that is undergoing either martensitic transformation or its reverse transformation. The state variables for the material are strain, temperature; extent of phase transformation, ξ, which is defined as the martensite fraction. The general state variable is defined as

$$\Lambda = (\bar{\varepsilon}, T, \xi) \tag{7}$$

The Helmholtz free energy is a function of the state variable Λ. The general constitutive relations can then be derived from the first and second laws of thermodynamics as

$$\bar{\sigma} = \rho_0 \frac{\partial \Phi}{\partial \bar{\varepsilon}} = \sigma(\bar{\varepsilon}, T, \xi) \tag{8}$$

The stress is a function of the martensite fraction, an internal variable. From the equation above, the rate for the mechanical constitutive equation is obtained as

$$\bar{\sigma} = \frac{\partial \sigma}{\partial \bar{\varepsilon}} \dot{\bar{\varepsilon}} + \frac{\partial \sigma}{\partial T} \dot{T} + \frac{\partial \sigma}{\partial \xi} \dot{\xi} = D \dot{\bar{\varepsilon}} + \Theta \dot{T} + \Omega \dot{\xi} \tag{9}$$

where D is Young's modulus, Θ the thermoelastic tensor; Ω the transformation tensor, a metallurgical quantity that represents the change of strain during phase transformation. The material properties derived from thermomechanics are given as

$$D = \rho_0 \frac{\partial^2 \Phi}{\partial \bar{\varepsilon}^2} \tag{10}$$

$$\Theta = \rho_0 \frac{\partial^2 \Phi}{\partial \bar{\varepsilon} \partial T} \tag{11}$$

$$D = \rho_0 \frac{\partial^2 \Phi}{\partial \bar{\varepsilon} \partial \xi} \tag{12}$$

If the expression of free energy is known, the minimization of the free energy may determine the equilibrium states of phases (i.e., the relation of the martensite fraction with the applied stress and temperature can be determined). However, instead of making the effort to find the free-energy expression, the martensite fraction, ξ, is assumed to be an exponential function of stress and temperature based on the study of transformation kinetics in Tanaka's model. These functions are

$$\xi_{M \to A} = \exp\left[A_a (T - A_s) + B_a \sigma\right] \tag{13}$$

$$\xi_{A \to M} = 1 - \exp\left[A_m (T - M_s) + B_m \sigma\right] \tag{14}$$

where A_a, A_m, B_a; B_m are material constants in terms of the transition temperatures, A_s, A_f, M_s; M_f.

Based on Tanaka's work, Liang and Rogers (1990) modified and extended this model to predict and describe quantitatively the behavior of shape-memory alloys. The above equation integrated with respect to time yields the constitutive equation

$$\sigma - \sigma_0 = D(\varepsilon - \varepsilon_0) + \theta(T - T_0) + \Omega(\xi - \xi_0) \tag{15}$$

A cosine model that better models the phase transformation was proposed

$$\xi_{A \to M} = \tfrac{1}{2}\left[\cos a_A (T - A_s) + 1\right] \tag{16}$$

$$\xi_{M \leftarrow A} = \tfrac{1}{2}\left[\cos a_M (T - M_s) + 1\right] \tag{17}$$

where the two material constants are determined from

$$a_A = \frac{\pi}{A_f - A_s} \tag{18}$$

$$a_M = \frac{\pi}{M_s - M_f} \tag{19}$$

The three transition temperatures, M_f, M_s; A_s are linearly dependent on the applied stress. To express the influence of stress on the transition temperatures, the following two material constants corresponding to the slope are introduced:

$$C_M = \tan \alpha \tag{20}$$

$$C_A = \tan \beta \tag{21}$$

where tan α and tan β are slopes. This effect of stress on the martensitic fraction versus temperature equations is introduced as follows:

$$\xi_{A \to M} = \tfrac{\xi_M}{2}\left\{\cos\left[a_A (T - A_s) + b_A \sigma\right] + 1\right\} \tag{22}$$

$$\xi_{M \leftarrow A} = \tfrac{1-\xi_A}{2}\left\{\cos\left[a_M (T - M_s) + b_{M\sigma}\right] + 1\right\} + \tfrac{1+\xi_A}{2} \tag{23}$$

where the two new material constants are

$$b_A = -\frac{a_A}{C_A} \tag{24}$$

$$b_M = -\frac{a_M}{C_M} \tag{25}$$

The constitutive equation, together with the martensitic volume fraction equations, provides the information necessary to solve all problems involving linear stiffness structures coupled to an SMA actuator.

Applications of Shape Memory Alloys – Shape-memory alloys have been used in many fields and applications over the past 30 years. The first major industrial application of a shape-memory alloy was a cryogenic pipefitting device developed by Raychem Corporation in 1969. Perhaps the best known example of this material is the prototype heat engine developed in the mid-1970's. Because of its biocompatibility and superior resistance to corrosion, Nitinol has been used in the medical field as a bone plate, an artificial joint; in dental applications. The major industrial application for this material has been as force actuators and robot controls (Funakubo, 1987).

Rogers and Robertshaw (1988) suggested the embodiment of SMA fibers into composites (known today as SMA hybrid composites) to adaptively control the performance of the composites or composite structures. The class of the materials referred to as SMA hybrid composites are composite materials with properties that can be changed or internal forces applied to it by activation of the embedded SMA wires (activation refers to changing the phase of the SMA; this is usually done through resistive heating). One of the many possible configurations of SMA hybrid composites is one in which the shape-memory alloy fibers are embedded in a material off the neutral axis on both sides of the structure in antagonistic pairs. SMA fibers can be embedded in a variety of matrix materials, such as graphite-epoxy, glass-epoxy, thermoplastic materials; other moldable or formable materials.

Two active structural modification techniques have been developed for SMA hybrid composites; one is called active properties tuning (APT), another is active strain energy tuning (ASET). In active properties tuning, the embedded fibers are actuated by resistive heating (i.e., passing an electrical current through the fibers). Once the temperature of the fibers exceeds the transition temperature, the Young's modulus of the SMA fibers may be increased in a controlled manner from 27 to 82 GPa. Active strain energy tuning uses the same mechanisms as described above for active property tuning but adds another important control parameter, the restoring stress associated with embedded inelastically elongated SMA fiber actuators trying regain their original shape by contracting.

Using either active properties-tuning or active strain energy tuning can result in significant control of the structural response of a composite structure. The difference between the two concepts is that APT changes the structural response by varying the stiffness of the structure alone, while ASET depends primarily on the recovery force of the plastically elongated fibers to change the structural response. Using active strain energy tuning in almost all cases results in more versatility of control than is possible in APT and can give a wide range of deflection, control. The ability of ASET in active vibration and structural acoustic control of SMA hybrid composites has been demonstrated both theoretically and experimentally.

Advantages and limitations of actuating materials – The main advantage of SMA materials is their capability to produce sizable actuation strains in the range of 4% to 6%. This feature makes the construction of demonstration experiments with SMA actuators very simple and accessible. Another advantage of the SMA materials is that they can be easily actuated through thermal effects. Since thermal effects can be readily obtained, e.g. through resistive heating of the SMA wire when an electric current is passed through, the practical realization of SMA based actuators can be easily attained. This feature reveals an outstanding advantage of SMA actuators, i.e. their inherent simplicity.

One main limitation of SMA actuators is their very poor energy conversion efficiency. This is usually due to the engineering of the SMA application. In the usual laboratory demonstrations, most of the heat generated through the passage of the electric current through the SMA wire is dissipated in the environment and does not produce any useful effect in the energy conversion process. For that reason, typical measured values for the energy conversion efficiency of SMA actuators do not exceed a few percent. These values could be certainly improved by proper thermal-flow management. However, this may happen at the expense of the attractive simplicity of the SMA actuators. At best, the energy conversion efficiency might be raise to the level of current thermal engines, which, by far, do not parallel the high efficiency of electric motors.

Self Repairing Materials

Compared to smart sensors and actuators, self-repair materials are in a much earlier stage of development. The only reported studies, from the University of Illinois, have demonstrated the feasibility of self-repair in both cementitious and polymeric materials. The concept of self-repair through the internal release of a chemical was first proposed by Dry (1990, 1991a). For a material to possess self-repair capability, it is necessary to have (i) a means for internal storage of the repair material (which is usually a polymer or a monomer) (ii) a stimulus to release the chemical; (iii) an approach to harden the chemical or dry out the water (if a monomer is used). Several alternatives have been suggested by Dry. In one approach, chemicals are stored inside hollow fibers. When cracking occurs to break the fiber, the chemical will be released. In another approach, a soft coating is placed around a hollow and porous fiber where the chemical is stored. When the fiber is pulled, surface sliding removes the soft coating, thus releasing the chemical. As an additional alternative, a wax coating can be put around a porous fiber containing monomers. On heating, the wax will melt and the monomer released. Further heating will then result in polymerization.

Experiments have been carried out to demonstrate the concept of self-repair. Wax-coated fibers containing methyl methacrylate were embedded in cementitious specimens. After a two stage heating process (at 50^0C and 100^0C), the specimens with the fibers show a much lower permeability than control specimens. SEM investigation has reviewed the release of methyl methacrylate into the cement matrix (Dry, 1991a). In another experiment, porous fibers containing calcium nitrite (a corrosion inhibitor) are covered with polyol. These fibers are placed in cementitious specimens that are exposed to salt water. Once the pH in the cementitious environment drops below 11.5, the polyol coating dissolves to release the calcium nitrite. Preliminary experimental results (Dry, 1991b) have indicated the potential of this approach in reducing the corrosion of steel reinforcing bars under high chloride concentrations.

White *et al.* (1997) from the University of Illinois have performed work on self repairing composites. The concept of self-repairing composites is based on the inclusion of polymeric microspheres that contain an uncured "healing" compound. When a crack advancing through the composite matrix hits the microsphere, the healing agent is released. It then flows into the crack to bridge the gap. Upon full curing of the released material, the crack is permanently bridged; the composite is considered repaired. They have focused their work on the repair of polyester resins by encapsulating a mixture of styrene, divinylbenzene (DVB), polystyrene and TBC (as inhibitor). Theoretical modeling (Hegeman, 1997) and preliminary experimental work (Jung, 1997) have shown that the microspheres can indeed be 'opened' by an advancing crack. Also, the encapsulated material can cure in-situ to bridge the crack.

Advantages and Limitations

The major advantage of self-repair material is to eliminate the need for regular inspection or monitoring, as the material can repair itself once damage occurs. However, the issues of multiple repairs and continuous self-repair capability (e.g., the replenishment of repair material with external supply) need to be studied further. Moreover, for repair mechanisms involving heating, the applications to large structures require additional investigations.

Current Status and Research Needs

The advantages and limitations of each type of smart materials have been summarized under the corresponding sections. As a summary, we can see that the field of smart materials is still in its infancy. For most of the smart materials discussed in this chapter, we have been able to demonstrate their 'smartness' and potential for structural applications in the laboratory. For real world applications, many important engineering issues, such as consistency, durability and cost, have not been addressed. Some smart materials have been used in other fields, such as mechanical or aerospace engineering. The direct transfer of such experience to civil engineering is very difficult, because we are under much tighter constraints. Specifically, civil engineers are dealing with large scale and very complex systems, but are often constrained by a limited budget. To promote the use of smart materials in civil engineering, considerable further research and development are required. Since the field of smart materials is so new, its research and development needs are many. A few that we believe to be most imperative include:

(1) The development of smart materials with better performance, together with higher durability and reliability

(2) The development of new applications that can take full advantage of material smartness

(3) The integration of various smart materials, or the combination of smart materials with conventional materials, to form smart material systems

(4) The reduction of production costs for smart materials and systems

(5) The development of smart materials and concepts applicable to large scale structural applications

Conclusion

In this chapter, we have looked at various materials that exhibit sensing, actuating or self-repair capabilities. For each of these smart materials, the physical principles behind the smartness are explained and practical applications described. The advantages and limitations of each material are also discussed. In summary, one can conclude that the potential of smart materials in structural applications has been clearly demonstrated through laboratory research in many institutions around the world. However, the field is still in its infancy and further research and development is required to establish smart materials as reliable, durable and cost-effective materials for large-scale civil engineering applications.

References

Optical Fiber Sensor

Alavie, A.; Maaskant, T. R.; Measures, R. M. (1993) "A Multiplexed Bragg Grating Fiber Laser System", *Photon. Tech. Lett.*, 5, 1112-1115.

Ansari, F.; Navalurkar, R. K. (1993) "Kinematics of Crack Formation in Cementitious Composites by Fiber Optics", *ASCE J. Eng. Mech.*, 119(5), 1048-1061.

Culshaw, B. (1984) *Optical Fibre Sensing and Signal Processing*, P. Peregrinus, London, UK

Culshaw, B. (1996) *Smart Structures and Materials*, Artech House, London, UK.

Davis, M.A. et al (1996) "Distributed Fiber Bragg Grating Strain Sensing in Reinforced Concrete Structural Components" *J. Cem. Conc. Comp.*, 1996.

de Vries, M. J.; Bhatia, V.; Claus, R. O.; Murphy, K. A; Tran, T.A. (1995) "Applications of Absolute EFPI Fiber Optic Sensing System for Measurement of Strain in Pre-Tensioned Tendons for Prestrained Concrete". *Smart Systems for Bridges, Structures and Highways*, SPIE 2446, 9-15.

Ezekiel, S. (Ed.) (1992) "Fiber Optic Gyros", *SPIE 15th Anniversary Conference*: 4-6 September 1991, Boston, MA.

Habel, W.; Hofmann, D. (1994) "Strain Measurements in Reinforced Concrete Walls during the Hydration Reaction by means of Embedded Fiber Interferometers", *Proceedings of 2nd European Conference on Smart Structures and Materials,* SPIE Vol. 2361, 176-179.

Habel, W.; Hillemeier, B. (1995) "Results in Monitoring and Assessment of Damages in Large Steel and Concrete Structures by means of Fiber Optic Sensors", *Smart Systems for Bridges, Structures and Highways*, SPIE Vol. 2446, 25-36.

Hale, K. F. (1992) "An Optical-Fiber Fatigue Crack Detection and Monitoring System", *Proceedings of 1st European Conference on Smart Structures and Materials*, Glasgow, 147-150.

Huston, D.; Fuhr, P.; Kajenski, P.; Snyder, D. (1992) "Concrete Bean Testing with Optical Fiber Sensors", *Nondestructive Testing of Concrete Elements and Structures*, edited by S. Sture and F. Ansari, 60-69.

Inaudi, D.; Elamari, A.; Pflug, L.; Gisin, N.; Breguet, J.; Vurpillot, S. (1994) "Low Coherence Deformation Sensors for the Monitoring of Civil Engineering Structures", *Sensor and Actuators A*, 44, 125-130.

Inaudi, D.; Vulliet, L, Pflug, L, Vurpillot, S.; Wyser, A. (1995) "Low Coherence Sensors for the Monitoring of Underground Works", *Smart Sensing, Processing and Instrumentation*, SPIE 2444, 170-178.

Inaudi, D.; Vurpillot, S.; Casanova, N. (1996) "Bridge Monitoring by Interferometric Deformation Sensors", *Laser Optoelectronics and Microphotonics: Fiber Optics Sensors*, SPIE, Beijing, Nov. 1996.

Kersey, A.D.; Morey, W. W.; "Multiplexed Bragg Grating Fiber Laser Strain Sensor System with Mode-Locked Interrogation", *Electronics Letters*, 29, 112-116

Leung, C. K. Y.; Elvin, N.; Olson, N.; Morse, T. F.; He, Y. (1997) "Optical Fiber Crack Sensor for Concrete Structures", *Smart Sensing, Processing and Instrumentation*, SPIE Vol. 3042, 283-292.

Lou, K.-A.; Yaniv, G.; Hardtmann, D.; Ma, G.; Zimmermann, B.D. (1995) "Fiber Optic Strain Monitoring of Bridge Column Retrofitted with Composite Jacket under Flexural Loads", *Smart Systems for Bridges, Structures and Highways*, SPIE 2446, 16-24.

Masri, Agbabian, Abdel-Ghaffar, Higazy, Claus and de Vries (1994) "Experimental Study of Embedded Fiber Optic Strain Gauges in Concrete Structures", *ASCE J. Eng. Mech.*, 120(8), 1696-1717

Measures, R. (1997) "Structural Monitoring with Fiber Optic Bragg Gratings" *Intelligent Civil Engineering Materials and Structures*, ASCE, pp. 180-201

Maaskant R. et al (1994) "Fiber Optic Bragg Grating Sensor Network Installed in a Concrete Road Bridge", *Smart Sensing, Processing and Instrumentation*, SPIE Vol. 2191, 457-465.

Michie, W. C.; McKenzie, I.; Culshaw, B.; Gardiner, P. (1994) "Optical Fibre Grout Flow Monitor for Post Tensioned reinforced Tendon Ducts", *Proceedings, 2nd European Conference on Smart Structures and Materials*, Glasglow, 1994.

Rossi, P.; LaMaou, F. (1989) *Materials and Structures* (RILEM), 22 (132), 437-442.

Tran, T.A.; Greene, J. A.; Miller, M.S.; Bhatia, V.; Murphy, K. A, Wang, A.; Claus, R. O. (1993) "Absolute Strain Measurements using the Extrinsic Fabry-Perot Interferometer", *FEORC Fiber Optics Review Conference*, Blacksburg, VA

Volanthen, M.; Geiger, H.; Dakin, J.P. "Distributed Grating Sensors using Low-Coherence Reflectometry", *J. Lightwave Technol.*; Vol. 15, No. 11, pp. 2076-2082 (1997)

Voss, K. F.; Wanser, K. H. (1994) "Fiber Sensors for Monitoring Structural Strain and Cracks", *Proceedings, 2nd European Conference on Smart Structures and Materials*, Glasglow, 144-147.

Wolff, R.; Miesseler, H. (1992) "Monitoring of Prestressed Concrete Structures with Optical Fiber Sensors", *Proceedings, 1st European Conference on Smart Structures and Materials*, Glasgow, 23-29

Piezoelectric Materials and Actuators

Crawley E. F.; Anderson, E. H. (1990) "Detailed Models of Piezoceramic Actuation of Beams", *J. Intell. Mater. Syst. Struct.* 1, 4-25.

EDO Corporation, *Piezoelectric Ceramics-Material Specifications, Typical Applications*. Salt Lake City, UT.

A. C. Eringen, *Continuum Physics*. Academic Press, New York, 1976.

Jaffe, B.; Roth, R. S.; Marzullo, S. (1954) "Piezoelectric Properties of Lead Zirconate-Lead Titanate Solid Solution Ceramics", *J. Appl. Phys.* 25, 809-810.

Liang, C.; Sun, F. P.; Rogers, C.A. (1994) "An Impedance Method for Dynamic Analysis of Active Material Systems", *ASME J. VB. Accost.* 116(1), 120-128.

Piezo Systems, Inc., *Piezoelectric Motor/Actuator Kit Manual*. Piezo Syst., Inc., Cambridge, MA.

Giurgiutiu, V.; Rogers, C.A. (1997) "Large-Amplitude Rotary Induced-Strain (LARIS) Actuator", *Journal of Intelligent Material Systems and Structures*, Technomic Pub. Co., Jan. 1997, pp. 41-50.

Electrostrictive Materials and Actuators

AVX Corporation, Electrostrictive Actuators. AVX Corp., Myrtle Beach, SC

D. Damjannovic and R. E. Newnham, (1997) "Electrostrictive and Piezoelectric Materials for Actuator Applications", *Journal of Intelligent Material Systems and Structures*, Technomic Pub. Co., Vol. 3 (1997) pp. 190-208.

K. Uchino. (1986) "Electrostrictive Actuators; Materials and Applications". *Ceram. Bull.* 65(4), 647-652 (1986)

Magnetostrictive Materials and Actuators

J. L. Butler (1988) *Application Manual for the Design of ETREMA Terfenol-D Magnetostrictive Transducers*, Tech.. Lit. ETREMA Products, Inc., Subsidiary of EDGE Technologies, Ames, IA, 1988.

Clark A. E. (1992) "High Power Rare Earth Magnetostrictive Materials", In *Proc. Recent Advances in Adaptive and Sensory Materials*, pp. 387-397. Technomic Publishing, Lancaster, PA, 1992. C. A. Rogers and R. C. Rogers Eds.

Goodfriend, M. J.; Shoop, K. M.; McMasters O. D. (1992) "Characteristics of the Magnetostrictive Alloy Terfenol-D Produced for the Manufacture of Devices",

Recent Advances in Adaptive and Sensory Materials, pp. 448-456. Technomic Publishing, 1992. C. A. Rogers and R. C. Rogers (Ed.).

Uchino K (1986) “Electrostrictive Actuators; Materials and Applications”. *Ceram. Bull.* 65(4), 647-652.

PVDF

Elf Atochem Sensors, Inc., *Standard and Custom Piezo Film Components*. Elf Atochem Sens., Inc., Valley Force, PA.

Lovinger, A. J. (1983) Ferroelectric Polymers. *Science*, 220, 1115-1121.

Electro-Mechanical Impedance Technique

Rogers, C. A.; Giurgiutiu, V. (1997) *Electro-Mechanical (E/M) Impedance Technique for Structural Health Monitoring and Nondestructive Evaluation*, Mechanical Engineering Department, University of South Carolina, OTT Disclosure ID No. 97162 of 7/2/97.

Giurgiutiu, V.; Rogers, C. A. (1996) “Electro-mechanical impedance technique for structural health monitoring and non-destructive evaluation” *Technical Report No. USC-ME-LAMSS-97-101*, Department of Mechanical Engineering, University of South Carolina, Columbia, SC 29208, January 1996.

Ayres, T.; Chaudhry Z.; Rogers C. A. (1996) "Localized Health Monitoring of Civil Infrastructure via Piezoelectric Actuator/Sensor Patches," *Proceedings, SPIE's 1996 Symposium on Smart Structures and Integrated Systems*, SPIE Vol. 2719, pp. 123-131.

Chaudhry, Z.; Lalande, F.; Ganino, A.; Rogers, C. A.; Chung J. (1995) "Monitoring the Integrity of Composite Patch Structural Repair via Piezoelectric Actuators/Sensors," *36th AIAA/ASME/ASCE/AHS/ASC SDM Conference*, New Orleans, LA, pp. 2243-2248.

Chaudhry, Z.; Sun, F. P.; Rogers, C. A. (1994) "Health Monitoring of Space Structures Using Impedance Measurements”, *Fifth International Conference on Adaptive Structures*, Sendai, Japan, 5-7 December, 1994; pp. 584-591.

Chaudhry, Z.; Joseph, T.; Sun, F.; Rogers, C. A. (1995) "Local-Area Health Monitoring of Aircraft via Piezoelectric Actuator/Sensor Patches", *Proceedings, SPIE North American Conference on Smart Structures and Materials*, SPIE Vol. 2443, pp. 268-276.

Childs, B.; Lalande, F.; Chaudhry; Z.; Rogers C. A. (1996) "High-Frequency Impedance Analysis for NDE of Complex Precision Parts," *Proceedings, SPIE's 1996 Symposium on Smart Structures and Integrated Systems*, SPIE Vol. 2717, pp. 237-243.

Esteban, J.; Lalande, F.; Rogers, C. A. (1996) "Theoretical Modeling of Wave Localization due to Material Damping," Proceedings, *SPIE's 1996 Symposium on Smart Structures and Integrated Systems*, San Diego, CA, 25-29 February 1996.

Giurgiutiu, V.; Chaudhry, Z.; Rogers, C.A. (1994) "The Analysis of Power Delivery Capability of Induced Strain Actuators for Dynamic Applications", *Proceedings of the Second International Conference on Intelligent Materials, ICIM'94*, June 5-8, 1994, Colonial Williamsburg, VA, Technomic Pub. Co., Inc., 1994, p.p. 565-576.

Keilers, C. H.; Chang, F.-K. (1995) "Identifying Delamination in Composite Beams Using Built-in Piezoelectrics: Part I - Experiments and Analysis; Part II An Identification Method", *Journal of Intelligent Material Systems and Structures*, Vol. 6, pp. 649-672, September, 1995.

Liang, C.; Sun, F. P.; Rogers, C. A. (1993) "An Impedance Method for Dynamic Analysis of Active Material Systems, *Proceedings, 34th AIAA/ASME/ASCE/AHS/ASC SDM Conference*, La Jolla, CA, 19-21 April 1993; pp. 3587-3599.

Tagged Smart Composites

Giurgiutiu, V.; Chen Z.; Lalande, F.; Rogers, C.A.; Quattrone, R. F.; Berman, J. B.(1996) "Passive and Active Tagging of Reinforced Composite Samples for In-Process and In-Field Non-Destructive Evaluation", *Journal of Intelligent Material Systems and Structures*, Technomic Pub. Co., Vol. 7, No. 6, Nov. 1996, pp. 623-634.

Giurgiutiu, V.; Jichi, F.; Quattrone, R. F.; Berman, J. B.(1999) "Experimental Study of Magnetostrictive Tagged Composite Strain Sensing Response for Structure Health Monitoring", *International Workshop of Structural Health Monitoring,* Sept. 8-10 1999, Stanford University, CA.

Quattrone, R. F.; Berman, J. B., White, S. R. (1998) "Self-Monitoring Structures Containing Magnetostrictive Composites", *Proceedings of the 21st Army Science Conference*, June 1998, Norfolk, VA, pp. 21-26.

Travillion, J.; Kamphaus, J.; Quattrone, R. F.; Berman, J. (1999) "Structural Integrity Monitoring Using Smart Magnetostrictive Composites", *Proceedings, 1999 International Composites Expo*, Cincinnati, OH, May 10-12, 1999, pp. 22D1 – 22D6.

White S. R.; Albers, R. G., (1996) "Magnetostrictive Tagging of Composite Materials for Structural Health Monitoring," USACERL Final Technical Report, Feb., 1996.

White, S. R.; Albers, R. G.; Quattrone, R. F., (1996) "Tagging of Composite Materials with Terfenol-D Magnetostrictive Particles for Structural Health Monitoring," ASME 1996 Mechanics and Materials Summer Meeting, June 12-14, 1996, Johns Hopkins Univ., Baltimore, MD.

White, S. R. (1999) "Magnetostrictive Tagging of Composites for Health Monitoring", *Proceedings, 1999 International Composites Expo*, Cincinnati, OH, May 10-12, 1999, pp. 22E1 – 22E6.

Self-Repairing Smart Composites

White, S. R.;. Sottos, N. R.; Geubelle, P. H.; Hegeman, A. J.; Jung, D. (1997) "Concepts for Producing Self-Repairing Polymeric Composites," USACERL Final Technical Report, July, 1997.

Jung, D.; Hegeman, A. J.; Sottos, N. R.; Geubelle, P. H.; White, S. R. (1997) "Self-Healing Composite Using Embedded Microspheres," ASME 1997 Int. Mech. Eng. Congress and Expo, Nov. 16-21, 1997, Dallas, TX.

Hegeman, A. J. (1997) "Self-repairing polymers: repair mechanisms and micromechanical modeling," M.S. Thesis, Dept. of Mechanical Engineering, Univ. of Illinois at Urbana-Champaign, Urbana, IL, June 1997.

Jung, D. (1997) "Performance and properties of embedded microspheres for self-repairing applications," M.S. Thesis, Dept. of Mechanical Engineering, Univ. of Illinois at Urbana-Champaign, Urbana, IL, July 1997.

Design of Induced-Strain Actuators

Giurgiutiu, V.; Rogers, C.A.; McNeil, S. (1997) "Static and Dynamic Testing of Large-Amplitude Rotary Induced-Strain (LARIS Mk 2) Actuator", *Journal of Intelligent Material Systems and Structures*, Technomic Pub. Co, Vol. 8, No. 7, June 1997, pp. 502-512.

Giurgiutiu, V.; Chaudhry, Z.; Rogers, C. A. (1997) " Design of Displacement Amplified Induced Strain Actuators for Maximum Energy Output ", ASME *Journal of Mechanical Design* Vol. 119, No. 4, December 1997, pp. 421-524.

Giurgiutiu, V.; Rogers, C. A. (1997) "Large-Amplitude Rotary Induced-Strain (LARIS) Actuator", *Journal of Intelligent Material Systems and Structures*, Technomic Pub. Co., Jan. 1997, pp. 41-50.

Giurgiutiu, V.; Rogers, C. A. (1996) "Power and Energy Capabilities of Commercially-Available Induced-Strain Actuators under Full-Stroke Dynamic Conditions", *Journal of Intelligent Material Systems and Structures*, Technomic Pub. Co., Vol. 7, No. 6, Nov. 1996, pp. 656-667.

Rusovici, R.; Giurgiutiu, V.; Rogers, C. A. (1996) "Experimental Study of Hydraulically-Amplified Induced Strain Actuator Proof-of-Concept Demonstrator", *Proceedings of the ASME Aerospace Division, AD-Vol. 52, ASME,* Nov. 1996, pp. 567-573.

Giurgiutiu, V.; Chaudhry, Z.; Rogers, C. A. (1996) "Energy-Based Comparison of Solid-State Induced-Strain Actuators ", *Journal of Intelligent Material Systems and Structures*, Vol. 7, No. 1, January 1996, Technomic Pub. Co., pp. 4-14.

Giurgiutiu, V.; Chaudhry, Z.; Rogers, C. A. (1995) "Engineering Feasibility of Induced-Strain Actuators for Rotor Blade Active Vibration Control", *Journal of Intelligent Material Systems and Structures*, Vol. 6, No. 5, September 1995, Technomic Pub. Co., pp. 583-597.

Shape Memory Alloy

Anderson, G. F.; Morrow, R. E. (1978) *Am. J. Orthod.*., 73(2) pp. 142-151.

Bank, R. (1975) *Shape Memory Effects in Alloys*, p. 537. Plenum, New York,.

Buehler, W. J.; Wiley, R. C. (1965) Nickel-Based Alloys. U.S. Pat. 3,174,851.

Castleman, L. S.; Motzkin, S. M.; Alicandri, F. P.; Bonawit, V. L. (1988) *J. Biomed. Mater.*. (10) pp. 695-731 (1976).[12] R. E. Newnham, The Golden Age of Electroceramics. *Adv. Ceram. Mater.* **3**(1) 12-16.

Cory, J. S. (1978) "Nitinol Thermodynamic State Surfaces" *J. Energy* **2**, No. 5.

Cross, W. B.; Kariotis, A. H.; Stimler, F. J. (1970) *Nitinol Characterization Study*. NASA. [contract Rep.] CR NASA CR-1433.

Funakubo, H. (1987) *Shape Memory Alloys*. Gordon & Breach, New York,.

Liang, C. (1990) "The Constitutive Modeling of Shape Memory Alloys", Ph.D. Dissertation, Virginia Polytechnic Institute and State University, Blacksburg.

Liang, C.; Rogers, C. A. (1990) "A One-Dimensional Thermomechanical Constitutive Relation of Shape Memory Materials", *J. Intell. Mater. Syst. Struct.* **1**(2), 207-234.

Tanaka, K.; R. Iwasaki (1985) "A Phenomenological Theory of Transformation Superplasticity", *Engineering Fracture Mechanics*, 21(4), pp. 709-720.

Muller, I. (1979) "A Model for a Body with Shape Memory", *Arch. Ration. Mech. Anal.* 70, 61-77.

Rogers, C. A.; Robertshaw, H. H. (1988) "Shape Memory Alloy Reinforced Composites", *Eng. Sci. Repr.* ESP25.88027.

Tanaka K.; Nagaki, S. (1982) "A Thermomechanical Description of Materials with Internal Variables in the Process of Phase Transformation", *Ing. Arch.* **51**, 287-299.

Sato and Tanaka (1988) "Estimation of Energy Dissipation in Alloys due to Stress-induced Martensitic Transformation", *Res. Mechanica*, Vol. 23, pp. 381-393, Elsevier Applied Science Publishers Ltd., U.K.

Roberto Lopez-Anido[1] and Vistasp M. Karbhari[2]

Chapter 2

Fiber Reinforced Composites in Civil Infrastructure

Abstract

Fiber reinforced polymer (FRP) matrix composites developed primarily for the aerospace and defense industries are a class of materials that present immense potential for use in civil infrastructure, both for the rehabilitation of existing structures and for the construction of new facilities. The results of research efforts in composites, initiated nearly a decade ago in several countries, are now being incorporated into civil infrastructure applications. Significant advances in the development of cost-effective reinforcing systems for concrete, wood and masonry, durable materials, reliable joining methods, optimized structural components and efficient repair and retrofit methods have been achieved. The objective of this chapter is to present the status of this class of emerging construction materials in civil infrastructure. Issues that influence the development of composites as structural materials for rehabilitation and new construction are analyzed in the context of specific applications. Examples of field implementation of FRP composites are presented and future needs and trends are discussed.

Keywords

Fiber, composite, FRP, polymer, laminate, pultrusion, retrofit, rehabilitation, structure, bridge, infrastructure

Introduction

Combining reinforcements and binder materials is a concept that has been extensively applied by designers to develop efficient construction materials. Beginning with the use of straw as reinforcement in mud and clay bricks by the ancient Egyptians, through the development of plywood as a form of laminated composite to take advantage of the direction of grains in veneer, "composite" material

---

[1]ASCE Member, Assistant Professor, Department of Civil and Environmental Engineering, University of Maine, 5711 Boardman Hall, Orono, ME 04469.

[2]ASCE Member, Associate Professor of Structural Engineering, Department of Structural Engineering, University of California, San Diego, 9500 Gilman Dr., La Jolla, CA 92093.

systems responded to the growing needs of building and transportation structures. For example, the combination of reinforcing steel and concrete, as a tailored macroscopic material system, has been the basis for a number of structural systems used for construction in the twentieth century. Civil engineers and designers continue to develop and adopt new forms of materials that would assist in the building of stronger, larger, more durable, energy efficient and aesthetic structures. Nature and the inevitably increasing demands of society have, however, also always increased the number of challenges facing the civil engineer. Such challenges result in the need for continuous development of enhanced materials and new material forms and systems that are capable of combining form and function, while ensuring that the structures built thereof fulfill the present and future needs of mankind.

At the beginning of the 21st century, several challenges and opportunities are facing civil engineers, foremost among them being the renewal of civil infrastructure facilities. The deterioration and functional deficiency of existing civil infrastructure represents one of the most significant challenges currently facing the nations of the world. In addition to deterioration caused through aging, weathering of materials and accidental damage to structures, traffic and housing/industrial needs have increased dramatically over the past few decades with transportation of goods and services being conducted on a global, rather than regional basis. There is a growing need for widening of highway systems to accommodate higher volume of traffic and for strengthening of existing structures to carry heavier loads at higher speeds. Enhanced understanding of structural response and mitigation of natural hazards such as earthquakes and storms has led to the establishment of new design codes and the consequent need to rehabilitate existing structures to ensure their continued safety.

Conventional materials such as timber, steel and concrete have a number of advantages, not the least of which is the relatively low cost of raw materials. However, it is clear that conventional materials and technologies, although suitable in many construction applications, and with a fairly successful history of past usage, lack in durability for demanding applications, and in some cases are susceptible to rapid deterioration, emphasizing the need for high-performance construction materials and systems. According to the Civil Engineering Research Foundation (CERF 1994), high-performance denotes one or more improved attributes, such as greater strength, improved durability and easier placement or fabrication, with their implication for reduce life cycle costs. Retrofit and rehabilitation of existing structures with conventional materials, in some cases, is not possible; with the traditional recourse being demolition and reconstruction. In all such (and other) cases, there is a critical need for the use of new and emerging materials and technologies, with the end goal of facilitating functionality and efficiency while increasing the overall durability and life span of the structures (See for example, Ballinger 1997).

Fiber reinforced polymer (FRP) matrix composites developed primarily for the aerospace and defense industries (See for example, Fletcher 1994) are a class of materials that present immense potential for use in civil infrastructure, both for the rehabilitation of existing structures and for the construction of new facilities. The objective of this chapter is to present the status of this class of emerging construction materials in civil infrastructure. Rather than presenting an overview of FRP composites in general, emphasis is given to their use in civil construction. Issues that influence the development of composites as structural materials for rehabilitation and new construction are analyzed in the context of specific applications. Examples of field implementation of FRP composites are presented and future needs and trends are discussed.

Fiber Reinforced Composites as a Class of Construction Materials

An Emerging Class of Construction Materials

For the purposes of this chapter, the term, fiber reinforced composites, will be restricted in its meaning to the combination of polymeric resins, acting as matrices or binders, with strong and stiff fiber assemblies which act as the reinforcing phase. The combination of the matrix phase with a reinforcing phase thus produces a new material system which conceptually is analogous to steel reinforced concrete, although the reinforcing fractions vary considerably (i.e., reinforced concrete in general rarely contains more than 5% reinforcement, whereas in FRP composites reinforcing volume fraction ranges from 35-65%), the matrix has both tensile and compression capacity and unlike concrete, polymeric resins used in this application area impart toughness to the overall FRP composite. Although FRP composites, in general, can be formulated for a variety of purposes, this discussion will restrict their functionality to the performance attributes related to mechanical, physical, chemical, thermal and electromagnetic properties only.

While the fibers are mainly responsible for strength and stiffness properties, the matrix contributes to load transfer and provides environmental protection. Typical fiber reinforcement materials are carbon, glass and aramid. Typical thermoset polymer matrices used for composites in infrastructure are polyesters, epoxies, vinyl esters and phenolics. A number of practical textbooks are available on composite materials. For example, mechanics of laminates are well presented by Jones (1999), materials and manufacturing aspects by Mallick (1993), hygrothermal effects and experimental methods by Daniel and Ishai (1994), laminate analysis by Reddy and Miravete (1995), and composites design by Tsai (1992) and by Barbero (1998).

The attractiveness of FRP composites as construction materials derives from a set of advantages gleaned from the tailorability of this material class through the synergistic combination of fibers in a polymeric resin matrix, wherein the fiber reinforcements carry load in predesigned directions (thereby taking advantage of anisotropy) and the resin acts as a medium to transfer stresses between adjoining fibers through adhesion and also provides protection for the fibers. This potentially gives the designer a wide palette of material choices to fit the specific requirements of the structure and shows immense potential to reinforce current materials being used in civil infrastructure.

Solutions for Civil Infrastructure Demands

In the search for material systems in civil infrastructure, fiber-reinforced composite materials can often offer a strong, durable, easy-to-install and lightweight solution as an alternative to conventional construction strategies. The results of a number of research efforts in composites, initiated nearly a decade ago in the U.S., Europe, Japan, Canada and other countries, are now being incorporated into several infrastructure applications (Mufti et al. 1991, Wilkins 1994, Busel 1995, FHWA 1997, Karbhari 1998). These applications of composites in infrastructure are leading to significant advances in the development of cost-effective reinforcing systems for internal and external reinforcement of concrete and prestressing cables and tendons, durable materials, reliable joining methods, optimized structural components and efficient repair methods (ACI 1996, Bassett 1998).

The predominant role of concrete as a construction material and the problems associated with corrosion of steel reinforcement fueled the development of fiber-composites for internal and external reinforcement of concrete and prestressing cables and tendons (ACI 1996, ACI 1998a, ACI 1998b). Furthermore, composites have also been successfully used for reinforcing wood (Plevris and Triantafillou 1992, Davalos et al. 1992, 1994, Gardner et al. 1994, Sonti and GangaRao 1996, Dagher et al. 1996, 1998, 1999, Tingley et al. 1997) and masonry (Priestly and Seible 1995, Ehsani et al. 1997).

Competition in the civil infrastructure market typically dictates high-volume low-cost materials with extended service-life and minimum maintenance. To attain these goals composites for construction differ from mainstream aerospace composites in material constituents, manufacturing techniques and joining methods. One example of this is the implementation of multi-axial heavy stitched E-glass fabrics for reinforcement of large structural components fabricated by pultrusion or vacuum assisted resin transfer molding (VARTM) (Howdyshell et al. 1998).

Advantages of FRP Composites over Conventional Construction Materials

Some of the most important advantages are listed and discussed below in the context of civil infrastructure applications:

a) High Specific Strength and Stiffness - Fiber reinforced composites show dramatic improvements in strength-to-weight ratios. For example, a comparison of typical ranges of FRP composite (glass and carbon fiber reinforced) characteristics with those of reinforcing steel can be noted from Table 1).

Fiber reinforced composites can also exhibit high stiffness-to-weight ratios as compared to most metals and alloys, as shown in Table 1. The combination of high specific strength and stiffness is a boon to designers, enabling them to develop designs at lower weights and thicknesses. Furthermore, these characteristics enable civil engineers to consider new design concepts (e.g. bridges that include longer free spans) that would be limited by the specific properties of other construction materials.

Table 1 - *Comparison of Typical Range of FRP Composite Properties with those of Mild Steel*

Property	Range	Comparison With Steel
Modulus	20 - 138 GPa	1/10 to 2/3 that of steel
Strength	340 - 1700 MPa	1 to 5 times the yield strength of mild steel
Strain to Failure	1 – 3 %	1/10 to 1/3 that of mild steel
Density	1.4 - 2.0 g/cm^3	4 to 6 times lighter than steel

In the civil infrastructure area the weight savings could result in enhancement in seismic resistance, increased speed of erection and a dramatic reduction in time for fabrication of large structures. Further, as in offshore applications, weight savings accrued in the superstructure translate to multiple levels of savings in the supporting and substructure elements. In the case of bridges this would translate to a decreased need for large foundations, whereas in buildings this would significantly decrease the size of columns in lower floors, enabling increased floor space and the construction of larger concert halls and auditoria unencumbered by clear span restrictions.

b) Enhanced Fatigue Life - Most composites are considered to be resistant to fatigue to the extent that fatigue may be neglected at the materials level in a number of structures, leading to design flexibility. This is considered to be a major advantage of carbon/epoxy composites in the aerospace arena as compared with metallic structures.

To characterize the fatigue behavior of structural materials an S-N (stress amplitude or maximum stress versus number of cycles) diagram is typically used, where the number of cycles to failure increases continually as the stress level is reduced. If below a certain value of stress no fatigue failure is observed then infinite material life can be assumed. The limit value of stress is called fatigue or endurance limit. For mild steel and a few other alloys, an endurance limit is observed at 10^5 to 10^6 cycles. For many FRP composites, an apparent endurance limit may not be obtained, although the slope of the S-N curve is substantially reduced at low stress level (Mallick 1993). In these cases, it is common design practice to specify the fatigue strength of the FRP material at very high number of cycles, e.g., 10^6 to 10^7 cycles, as the endurance limit. Unlike metals, FRP composites subjected to cyclic loads can exhibit gradual softening or loss in stiffness due to microscopic damage before any visible crack appears. For example, potential fatigue damage mechanisms in unidirectional fiber reinforced composites loaded parallel to the fibers are: fiber breakage, interfacial debonding, matrix cracking and interfacial shear failure (Talreja 1987).

Damage and cracking resulting from fatigue and fretting fatigue is one of the reasons for significant distress in bridge and building components. A decreased concern related to fatigue resulting from the use of composites can lead to significant innovation in structural design, especially in seismic areas, or in building structures housing large vibrating machinery. For example, bridge decks made of E-glass/vinyl ester composites fabricated by pultrusion and by vacuum assisted resin transfer molding (VARTM) (Lopez-Anido et al. 1998) and FRP-concrete hybrid decks (Lopez-Anido et al. 1999) have not shown damage accumulation during fatigue tests up to two million load cycles. However, the fatigue resistance of bonded and bolted connections may control the life of the structure. For structural applications where the design is typically stiffness driven (e.g., FRP bridge decks), working stresses are generally very low compared with ultimate strength (See for example Harik et al. 1999a, 1999b, 1999c, 2000) resulting in an extended fatigue life.

c) Corrosion Resistance - Unlike metals, composites do not rust, making them attractive in application where corrosion is a concern. This has led to an intense interest in composite reinforcing bars and grids, as well as cables for pre- and post-tensioning, and for use in cable stays. A report from BIRL/Northwestern University (1995) identified corrosion mitigation as one of the main reasons for application of composite materials in infrastructure. According to this report composite material components should be used in applications where standard metallic components incur high maintenance costs due to corrosion and its effects. An example in bridges is redirecting deicing salt run-off with a corrosion-resistant drainage system. Other areas where composites have been applied for corrosion resistance are: marine waterfront structures, cladding panels, pipelines, mining and walkways in harsh environments (Busel 1995).

d) Controllable Thermal Properties - In a number of applications, temperature gradients and temperature induced expansion and contractions result in thermal strains that may be of concern to civil designers. Thermal gradients through the depth of a steel box-girder bridge, for example, have resulted in severe distress and occasional failures in areas having extreme heat and temperature fluctuations. While steel has isotropic properties, FRP composites exhibit different coefficients of thermal expansion in the fiber direction and perpendicular to the fiber direction. See for example Carbon FRP properties in Table 2 for a fiber volume fraction of 60%. The low orthotropic thermal conductivity of FRP composites means that they are typically good thermal insulators. Therefore, control of thermal strains in FRP composite structures needs to account for orthotropic thermal properties in the design of the fiber reinforcement layup. For example, in the design of FRP composite bridge decks it is necessary to match orthotropic coefficients of thermal expansion of the composite layup with other attached structural and non-structural components (e.g., supporting beams, wearing surfaces, core materials).

Table 2 - *Comparison of Representative Properties of Reinforcing Steel, Concrete, Carbon Fibers, and Carbon FRP Composites ($V_f = 60\%$)*

Property	Units	Concrete	Grade 60 Steel	Carbon Fiber	Carbon FRP Composite	
					In the Fiber Direction	90° to Fiber Direction
Density	(g/cm^3)	1.61	7.8	1.75-1.80	1.6	1.6
Tensile Modulus	(GPa)	2.5	200	230	160	10
Tensile Strength	(MPa)	3.5	400 (Yield)	3000	1725	40
Elongation to Break	(%)	---	0.2 (Yield)	0.9 - 1.5	1.1	1.5
Poisson's Ratio	---	0.2	0.3	0.2	0.22-0.28	0.015-0.02
Thermal Conductivity	($W\ m^{-1}°K^{-1}$)	1	50	7.0 - 8.5	11-18	2 - 3
Coefficient of Thermal Expansion	($10^{-6}/°C$)	10	12	-0.4	0.02-0.04	20

e) Parts Integration - With most construction materials, large structural systems have to be made in sections requiring secondary joining operations in the field that not only increase costs, but also decrease overall system reliability. An example of this is a large truss or frame structure, wherein the assembly of a large number of components can result in cost increases over the hypothetical cost of fabricating the structure in one operation. The questions related to the reliability and integrity of field welding associated with stiffeners for girders, or that in steel shells/casing used in seismic column retrofit are similar.

With fiber-reinforced composites, it may be possible to fabricate large, complex parts in one operation using insitu-processing methods such as resin infusion, reducing the number of joining operations and increasing overall safety. An example of parts integration in the transportation industry is composite buses that utilize a single piece structure containing the vehicle's body and chassis elements. The material utilized in this novel, unitized structure consists of E-glass/vinyl ester composite laminates, produced using a patented resin infusion molding process known as SCRIMP® technology.

f) Tailored Properties - Metals and most building materials intrinsically force the use of structural designs that are isotropic and hence inefficient irrespective of whether there is a need for similar properties in all directions. For example, the seismic retrofit of concrete columns requires that the shell/casing provide additional hoop reinforcement in order to develop confinement. The use of steel results in additional strength and stiffness both in the hoop and axial directions, with the additional axial stiffness often causing further distress due to the attraction of forces during a seismic event to the stiffest axial member. In contrast, with composites it is possible to tailor properties to comply only in the directions required, thereby improving efficiency and economy (Seible and Karbhari 1997). Representative properties of reinforcing steel, concrete, carbon fibers, and orthotropic carbon fiber reinforced composites can be compared on the basis of Table 2.

g) Non-Magnetic Properties - Steel members and reinforcements are often a problem in structures such as hospital operating theaters, areas where radar is operative/housed, and areas which house antennae and other sensitive electronic equipment due to their propensity to cause interference under electromagnetic waves. The use of glass fiber reinforced composite components is often a solution to these problems.

h) Lower Life-Cycle Costs - Due to the corrosion resistance and enhanced resistance to solvents and the environment, composite structures are expected to require less maintenance, resulting in lower overall life-cycle costs (Ehelen and Marshall 1996). Anecdotal evidence provides substantial reason to believe that, if appropriately designed and fabricated, FRP composite structures have extended longevity, resulting in low life-cycle costs. However, actual comparative long-term data is sparse, not well documented, or not easily accessible to the civil engineer. Furthermore, available data on durability can be contradictory (e.g., different forms of material and different processes) and hence can create uncertainty (Karbhari et al. 2000).

i) Faster Field Installation - Civil construction is often characterized by long construction and installation periods, which can result not only in undue but hitherto uncorrectable delays in the opening of facilities but also in considerable inconvenience to users (such as in the case of road diversions, lane blockages, and posting of speed limits, related to repair or even extension of current roads and bridges). Further, construction using conventional materials is often seasonal, resulting in prolonged periods wherein no work is possible. In contrast, large composite parts can be fabricated off-site or in factories due to their light weight and can be shipped to the construction site and installed using light (rather than heavy and specialized) equipment, thereby minimizing the amount of site work. This can lead to year-round installation of composite structures with its attendant increase in overall construction efficiency and positive effect on planning and logistics. However, field joining of composite structural components may require further developments in adhesive bonding under varying pressure, temperature and moisture conditions.

Challenges in the Adoption of FRP Composites as Construction Materials

Notwithstanding the considerable advantages and opportunities afforded by FRP Composites, civil engineers should note that there still exist significant challenges that must be overcome before this class of emerging materials can be used as widely as conventional materials such as timber, concrete and steel. Some of the major challenges are presented in this section. However, it should be noted that in many cases these challenges relate to intangibles that may be resolved through education, research, implementation and better exposure through projects.

a) Higher Initial Material Cost – Higher material cost is perceived as the major obstacle for using composites in an extremely cost-competitive construction market. The constituent materials in FRP composites (e.g., fiber reinforcement and resin matrix) are more expensive on a equivalent mass basis than conventional construction materials. However, composite materials weigh significantly less and, as elucidated in the previous section, have significantly higher strength-to-weight and stiffness-to-weight ratios enabling the same function to be derived through the use of less material. Therefore, cost analysis of composites with respect to conventional materials should be based on a per unit performance, with proper consideration to weight reductions.

The overall cost of the FRP composite includes also the manufacturing cost. These costs can be reduced with a continuous fabrication process that minimizes labor, such as pultrusion. Alternatively, flexible fabrication methods for large structural components that do not require expensive tooling, such as vacuum assisted resin transfer molding (VARTM), can lower as-fabricated costs.

b) Lack of Familiarity with Materials - Elastic and strength properties of composite materials vary depending on the material constituents (fiber, matrix and fillers), fiber architecture and fiber volume fraction (See for example Mallick 1993). Besides five independent elastic constants are needed to characterize a composite assuming that the material is transversely isotropic (Barbero 1998). This is a major difference between fibrous composites and conventional isotropic materials such as steel, which have been standardized through the use of accepted grade types. In this sense, the tailorability of composites may be both a designer's dream and a nightmare.

FRP composites further can fail through different mechanisms depending on microstructure and fiber architecture (e.g., fiber breakage, matrix cracking, debonding, fiber pullout), making it difficult to devise a good structural design. In addition, a laminated structure can exhibit planes of weakness between layers, as well as local variations in properties due to resin-rich areas, voids and fiber misalignments.

Nevertheless, civil engineers can become familiar with composites by learning and adapting standards and guidelines for material characterization, which are used by other industries. For example, the ASTM (2000) Committee D-30 has compiled useful standards for the characterization of high modulus fibers and composite materials. These standards cover methods for testing constituent properties, thermal and physical properties, lamina and laminate mechanical properties and structural properties. The scope of application of the methods ranges from quality control through design data generation. SACMA (1996) has compiled standards for safe handling of composite materials, as well as, acceptance requirements of various composite materials. As defense conversion and dual use technology programs encourage the diversification of composite applications for civil infrastructure and other commercial uses, new standard development is required to support the associated materials characterization needs.

c) Lack of Knowledge on Design of Composite Structures - Engineers typically find design with composites more involved than with conventional materials because one is required to simultaneously design both the material itself and the geometry of the part. The lack of standardized grades of material, as well as the inability to reshape from stock shapes and profiles (as is usual with steel bars, plates, and profiles) can prove to be a major hindrance in using composites by engineers schooled in the traditional use of construction materials. Design and fabrication, which previously could be separate functions must now be considered concurrently and often by the same individual.

Nonetheless, civil engineers can overcome these difficulties by learning and adapting the methodology for analysis and design of composite structures, which is used by other industries (See for example Composite Materials Handbook-MIL-17 1999). This requires specifying experimental procedures for material characterization and understanding their correlation with the elastic and strength properties used for design (Daniel and Ishai 1994). It also entails using laminate analysis to predict hygro-thermal and mechanical strains and stresses (Jones 1999). Laminate analysis is also a practical approach for computing stiffness properties of laminated structural components. Designing efficient layups can be systematically attempted with practical design methods (e.g., use of carpet plots). Finally, understanding of modes of failure facilitates selecting and applying appropriate failure criteria for laminate strength design (Barbero 1998).

d) Lack of Comprehensive Standards and Design Guidelines - Civil design and construction is widely dominated by the use of codes and standards predicated on the use of well-documented and standardized material types. The intrinsic tailorability of composites through selection of constituent materials and reinforcement architectures can be considered as a challenge in adoption if mere standards rather than performance based standards (which are only now being developed and accepted even for conventional methods and materials of construction!) are followed. A number of efforts are ongoing to develop civil/structural engineering design specifications for FRP composites. For example, the need to develop national standards on the use of FRP composites for the rehabilitation of concrete and masonry structures has been recently addressed by the National Institute of Standards and Technology (Duthinh 1999).

ASCE (1984) pioneered a manual to provide guidelines to structural engineers engaged in the design of plastics and reinforced plastic structural components. Most recently the EUROCOMP design code and handbook (Clarke 1997) was compiled as a practical tool for the European construction industry. The scope of this code is limited to glass FRP materials, components, connections and assemblies for civil structures. The American Concrete Institute Committee 440 has drafted design recommendations for concrete reinforced with FRP bars (ACI 1998a), guidelines for selection, design, installation of FRP systems for externally strengthening concrete structures (ACI 1998b), and standard test methods for FRP rod and sheet (ACI 1998c). Acceptance criteria for fiber-reinforced composite systems used to strengthen concrete and masonry structural elements was issued by ICBO (1997).

Recent work on developing specifications for fiber-reinforced composites has focused on probability-based Load and Resistance Factor Design (LRFD). The main advantages of using LRFD are (Ellingwood 2000); a) Uniform basis for design with all construction technologies (e.g., concrete, steel, wood); b) Enhances confidence in new building products (controls reliability); c) Facilitates implementation of research in standards; d) Eliminates unnecessary conservatism in design (more economical design) and e) Encourages new technologies and products. Along these lines, Mertz and Kulicki (1997) presented a discussion on the extension of AASHTO LRFD Bridge Design Specifications to FRP composite materials bridges.

e) Need for Integrated Materials-Process-Design Structure - As mentioned previously, efficiency of FRP composites, and components or structures derived thereof, is largely dependent on the integration of facets of materials selection, process methodology and overall design. These aspects, although known in conventional construction, are rarely used together, and hence the need for their use in FRP composites, can present a major obstacle for acceptance by the civil engineering community. An example of integration of materials, process and structural design has been the recent advances in pultrusion and VARTM processing through the use of multiaxial fabric reinforcements and toughened resin matrices resulting in optimized designs for structural components (e.g. bridge decks, piles, linear profiles). However, these advances require that civil engineers interact with other professionals in the areas of polymer chemistry, composites processing, fiber reinforcements, and material characterization.

f) Lack of Knowledge on Connections - The design of connections in FRP composite structural systems is still in its infancy with designs being adopted at present from metallic analogues rather than being developed for the specific performance attributes and failure modes of FRP composites. This has often resulted in the use of high margins of safety causing designs to be cost inefficient, or leading to premature failure. Critical connection problems associated with application of composites in construction include issues of attachments, flexible joints, and field connections. In general, joints and connections should be simple, durable, and efficient to provide adequate deformability. Similar to other construction technologies, the connections should not form the weak link in the overall system (Nanni et al. 1996).

A wealth of information is available from the aerospace industry on joints, splices, and connections of FRP composites (Banuk 1996, Composite Materials Handbook-MIL-17 1999, Ramkumar et al. 1986, Hart-Smith 1982), but only limited use has been made of this resource, perhaps due to the inherent disconnect between aerospace and civil design methodologies. Bolted, bonded and combined connections have been successfully developed and utilized for composite structures (See for example Bank et al. 1993, Bank et al. 1996, Crasto et al. 1996, Mottram and Zheng 1996, Turvey 1997). A comparison of joining techniques for FRP composites is presented in Table 3. Most recently, interlocking connections have been developed to optimize integration of composite components into an overall structural system, i.e., pultruded modular bridge decks (Lopez-Anido and GangaRao 1997).

g) Uncertain Durability - Although polymeric matrices are susceptible to degradation in the presence of moisture, temperature and corrosive chemical environments, the main concern related to the durability of FRP composites is the lack of substantiated data related to their long-term durability. It should be kept in mind that FRP composites have only been used, even in the aerospace world, for structural components for little over 50 years, and therefore there is even lack of substantial anecdotal evidence. Further the resin systems and manufacturing methods that are likely to be used in civil infrastructure applications are not the same as those that have been characterized in the past by the US Air Force and the aerospace industries, rendering the extensive databases on materials such as AS4/3501-6 and T300/5208 little more than references for current materials systems.

To fully utilize the potential of composite materials their durability must be known. A comprehensive review on durability of FRP composite materials for infrastructure applications was compiled by Liao et al. (1998). A complete understanding of the degradation mechanisms under various environmental exposures and loading conditions will facilitate the development of accelerated test methods for service life prediction (Chin et al. 1997, Lesko et al. 1997, Gentry et al. 1998).

Key durability concerns for fiber-reinforced composites in civil infrastructure have been identified by a panel coordinated by NIST and CERF (Karbhari et al. 2000), as follows: a) Moisture and chemical reagents; b) Alkaline solutions; c) Thermal effects; d) Creep rupture; e) Fatigue resistance; f) UV weathering; and g)

Fire performance. Synergistic effects (i.e., combination of stress, moisture, chemicals and temperature effects) are extremely important and need to be assessed. Another important aspect is the effect of size (Zweben 1994) on durability performance, considering that most durability studies are conducted with coupon specimens, while the dimensions of FRP structures are much larger.

Table 3 – *Comparison of FRP Composite Joining Techniques (Adapted from the Structural Plastics Design Manual, ASCE 1984)*

Issue	Mechanical Fastener	Adhesive Bond
Stress concentration at joint	high	medium
Strength/weight ratio	low	medium
Seal (water tightness)	no	yes
Thermal insulation	no	yes
Aesthetics (smooth joints)	bad	good
Fatigue endurance	bad	good
Sensitivity to peel loading	no	yes
Disassembly	possible	impossible
Inspection	easy	difficult
Skill required of fabricator	low	high
Heat or pressure required	no	yes
Tooling cost	low	high
Time to develop strength	immediate	long

The path forward proposed by the NIST-CERF panel (Karbhari 2000) is based on three actions: 1) Establish and maintain a durability database; 2) Prioritize research on critical durability areas according to a unified protocol for testing, data collection and validation, and 3) Assess periodically the state of composites in the field. To attack the problem of assessing environmental effects in composite materials, civil engineers can follow the methodology proposed by Springer (1988), as shown in Figure 1.

h) Perception of Brittle Material Response - FRP composites do not show definite yield like steels. Thus the lack of ductility at the materials level can be a cause of concern to some designers. However, at the structural level, components fabricated from FRP composites can be designed to exhibit a sequence of damage mechanisms, which ensures a relatively slow failure with extensive deformation, leading to a progressive and safe mode of failure. One example of a structural system that can develop extensive deformation prior to failure is FRP bridge decks (Karbhari et al. 1997a, Lopez-Anido et al. 1998, Harik et al. 1999a, 1999b, 1999c, 2000).

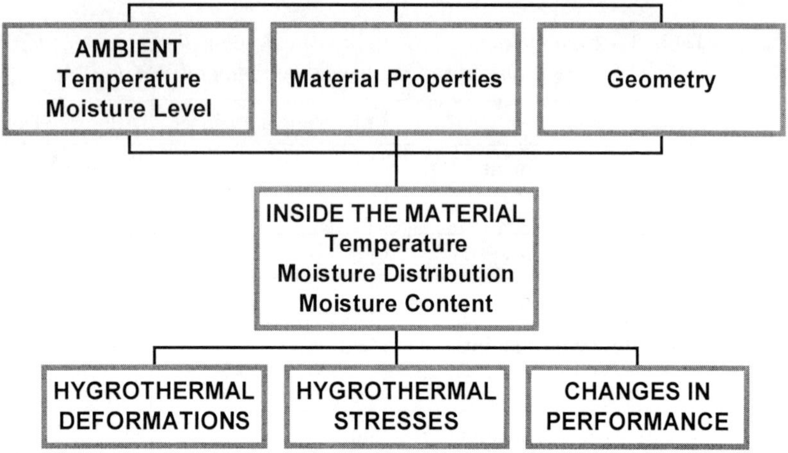

Figure 1 - *Procedure for Assessing Environmental Effects (Springer 1988)*

However, FRP composites can contribute to increase the ductility of other structural systems. For example, the rehabilitation of reinforced concrete columns jacketed with fiber-reinforced composites cannot only improve shear strength and confinement, but also the ductility of the system (Saadatmanesh et al. 1994, Seible and Karbhari 1997, Gergely et al. 2000). Another example is the reinforcement of glued-laminated wood beams with FRP composite tension reinforcement. In this case, properly reinforced FRP-glulam beams not only exhibit significant strength increases, but also they can develop wood ductile compression failure, rather than the typical brittle tension failure of wood (Dagher et al. 1996, 1998).

Materials Perspective for FRP Composites Design

Due to the combination of strong and stiff fibers (Table 4) with polymeric resin matrices, FRP composites have unique attributes and have to be designed keeping in mind specific characteristics of anisotropy, inhomogeneity and laminated or layered structure (Mallick 1993, Daniel and Ishai 1994, Barbero 1998, Jones 1999). Other material considerations in designing FRP composites are non-ductile unidirectional response, damage mechanisms, coefficients of thermal expansion, internal damping and corrosion resistance.

It should be emphasized that the properties of FRP composites depend on the properties of material constituents (i.e., reinforcing fiber, matrix, fillers) and the corresponding volume fractions. Thus, although the modulus of a carbon fiber could be in the 230 GPa range (T-300 from Table 4), and hence greater than that of Grade 60 steel at 200 GPa, the modulus of an unidirectional carbon fiber reinforced composite is actually only in the range of 160 GPa in the direction of the reinforcement and only 10 GPa in a direction transverse to the fiber direction for a fiber volume fraction of 60% (See Table 2).

Table 4 - *Properties of Typical Reinforcing Fibers*

Fiber	Designation	Density (g/cm^3)	Tensile Modulus (GPa)	Tensile Strength (MPa)	Strain to Failure (%)	Tow Count
Glass	E-Glass	2.55	72.4	3450	4.8	---
	S-Glass	2.49	86.9	4800	5.2	---
Pan Carbon	T-300[a,b]	1.76	230	3530	1.5	1, 3, 6, 12 K
	AS4[c]	1.80	221	3930	1.7	3, 6, 12 K
	IM6[c]	1.74	275	5240	1.7	12 K
	UHM[c]	1.85	441	3445	0.8	3, 6, 12 K
Pitch Carbon	P-55[a]	2.0	380	1900	0.5	---
	P-100[a]	2.15	758	2410	0.32	---
Aramid	Kevlar 49[d]	1.45	131	3620	2.8	---
	Technora[g]	1.39	74	3500	4.6	---

[a]BP/Amoco, [b]Toray, [c]Hexcel, [d]DuPont, [g]Teijin

The choice of reinforcement and polymer matrix should be made on the basis of the structural application, surrounding environment, lifetime requirements, and manufacturing process to be used (See for example Seible et al 1997, Karbhari and Seible 1997, and Karbhari and Seible 1998). Although there is a tendency in the civil infrastructure community to choose one fiber type predominantly (e.g., carbon over glass, or glass over carbon), it is emphasized that both types have advantages and disadvantages, and hence their choice must be dictated by the details and requirements of the application under consideration.

Carbon fibers are available in a range of moduli, with some grades even achieving moduli as high as 3 to 4 times that of steel (See Table 4). However, in all these varieties strain capacity is severely limited, whereas glass fibers have much greater strain capacity, and hence greater ability for being used in designs that incorporate modes of noncatastrophic failure. Carbon fibers are by chemical structure intrinsically inert to most environmental influences and hence may be used safely in applications wherein severe environmental exposure is expected. The environmental sensitivity of glass fibers depends on the type of fiber used (e.g., E-, S-, AR-, etc.), but in general the fibers are attacked in the bare state by alkalis and moisture, and are also susceptible to stress corrosion cracking. Aramid fibers have a tendency to absorb water, are susceptible to significant creep, and are nonlinear in compression failing due to kinking and localized buckling after fibrillation.

Although a range of resin systems is available to the designer, their selection must be based on the exigencies of processing (e.g., viscosity, cure temperature and profile, moisture sensitivity), fiber-matrix compatibility (sizing or fiber surface treatment), and durability performance required (e.g., in-service temperature, toughness, environmental resistance – moisture absorption, UV weathering resistance, fire resistance).

The uncertainties in the material properties of FRP composites, including the long-term performance, can be accounted in design by considering partial safety factors or knock-down factors (Clarke 1997, Karbhari et al. 1998). Uncertainties can be related to the following causes: a) Level of uncertainty based on the derivation of material properties from test values or models; b) Characteristics of the fabrication process and degree of cure; and c) Operating temperature and load duration. It is worth noting that variability of elastic and strength properties is typically higher for matrix-dominated properties (e.g., transverse tensile strength of a unidirectional lamina) than for fiber-dominated properties (e.g., longitudinal tensile strength of a unidirectional lamina).

Applications to the Rehabilitation of Structures

Composites applications in infrastructure can be classified into the general areas of structural rehabilitation and new construction, as shown in Figure 2. Generically, FRP composites can be utilized for structural rehabilitation in the following situations: a) Corrosion arising from the use of de-icing salts or proximity to harsh environments; b) Errors in design and construction; c) Modifications made to a structure and changes of use; d) Need for blast-proofing industrial plants or buildings; e) More stringent design criteria such as for seismic resistance; f) Aging of materials such as in historical structures; g) Fire damage; and h) Vehicle impact.

Structural rehabilitation with composites offers several advantages over using composites for new construction. Through rehabilitation (i.e., repair, strengthening or seismic retrofit) of constructed facilities with fiber-reinforced composites, the civil engineer will: a) Reduce the application risk of a new material technology; b) Require fewer design and building code provisions; c) Facilitate construction permits; and d) Compete favorably with conventional materials and methods.

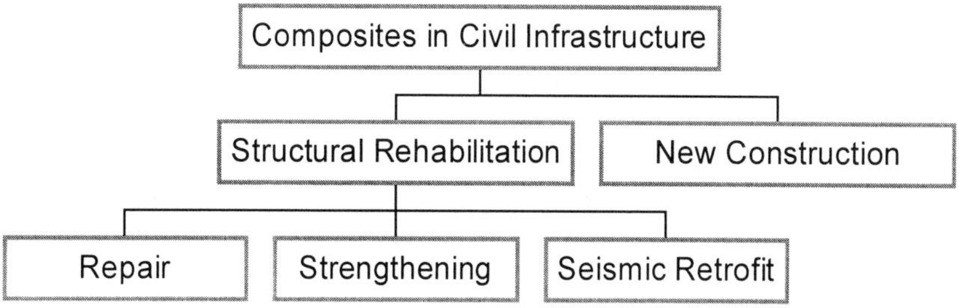

Figure 2 - *Classification of Composites in Civil Infrastructure*

In the area of new construction with fiber-reinforced composites four main applications have been identified (Nanni et al. 1996): a) Reinforcement for concrete with rebars, bonded plates, jackets or fabrics; b) Prestressed reinforcement for concrete with tendons and cables; c) Structural shapes used for beams, columns and decks; and d) Structural systems, such as composite shells for columns. Within the area of structural rehabilitation of concrete, four main applications have been identified in Table 5. Each of these applications requires tailoring the fiber-reinforced composite system to meet structural and material requirements.

Table 5 - *Rehabilitation and Retrofit of Concrete Structures with Fiber-Reinforced Composites*

Application	Description
I-girder and box beam repair / retrofit	Design for repair: impact damage and other in-service damage, including corrosion-induced damage; Design for strengthening: Flexural, shear and torsional live load upgrading, or rehabilitation owing to inadequate original design.
Seismic column retrofit	Design for specific modes of failure: shear, flexural hinge confinement, lap splice clamping, and bar buckling restraint.
Deck soffit applications	Design for crack repair and live load upgrade.
Substructure repairs	Design for repair and strengthening: pile and column bent caps, pier walls and non-seismic column.

The rehabilitation and retrofit of existing concrete structures with polymer matrix composites can be generally accomplished in one of two ways: a) Application of composite overlays or strips, and b) External post-tensioning using composite cables, tendons or bars. The use of composite overlays in the form of complete coatings or in the forms of strips or patches is applicable to the following cases: a) Seismic retrofit and repair of bridge columns; b) Strengthening of bridge superstructure (deck soffits and girders) for increased capacity; and c) Seismic retrofit and strengthening of reinforced and unreinforced concrete and masonry walls.

A variety of fiber-composite bonded systems for external repairing and retrofitting concrete members have been developed, as shown in Table 6. The application of these systems raises specific materials issues and concerns that need to be considered by civil engineers.

Rebars

The design protocols available in current codes for reinforced concrete structures are based on a combination of engineering principles and empirical equations developed through extensive experimental work and many years of experience (ACI 1998a). Unfortunately, current codes are applicable only to steel reinforcement and provide no guidance on how to modify the existing requirements to use FRP composites for concrete reinforcement. Countries, such as Japan (JSCE 1997) and Canada (CSA 1996), have already established design procedures specifically for using FRP for concrete reinforcement. Extensive testing and analysis of FRP materials for concrete reinforcement were conducted in Europe resulting in a number of bridge demonstration projects (Taerwe 1997). FRP rods were used to reinforce a concrete deck on steel stringers in the McKinleyville Bridge, West Virginia (GangaRao et al., 1997). Most recently the ACI Committee 440 (ACI 1998a), has drafted design recommendations for concrete reinforced with FRP bars.

FRP reinforcement products for concrete are made of continuous fibers of aramid (AFRP), carbon (CFRP), or glass (GFRP) impregnated in a polymer matrix (ACI 1996). Several FRP reinforcement products are available in the form of grids, rods, fabrics and ropes. For an effective reinforcing action, it is necessary to develop bond strength between FRP and concrete. This is attained in FRP rods by having various types of deformation systems, including exterior wound fibers, sand coatings and separately formed deformations (ACI 1998a). However, the bond strength of deformed FRP rods is in general lower than that of equal diameter steel bars (Nanni 1998). The advantages and disadvantages of FRP reinforcement over conventional steel reinforcement for concrete discussed by (Erki and Rizkalla 1998) are summarized in Table 7.

Unlike steel reinforced concrete, FRP reinforced concrete cross sections have to be over reinforced (Nanni 1998). The failure mode controlled by the tensile rupture of the FRP bars is catastrophic and therefore undesirable. Hence, a minimum amount of FRP reinforcement needs to be imposed to prevent this mode of failure. According to (ACI 1998a) the design should be such that concrete crushes in compression before the tensile rupture of FRP reinforcement occurs. To guarantee this the FRP reinforcement ratio should be larger than the ratio at balanced failure (i.e., reinforcement ration producing a condition of simultaneous failure for both concrete and FRP).

Table 6- *FRP Composite Repair and Retrofit Systems for Concrete Structures*

System	Aspects	Issues / Concerns
Wet Lay-up of fabric	• Use of a wet bath or impregnator • Manual or semi-automated process • Ambient cure • Flexible • Ease of use in restricted areas	• Quality control of mix • Wet-out • Non-Uniform resin distribution • Compaction & fiber wrinkling • Control of cure • Environmental issues
Wet Winding of Tows	• Use of dry tow impregnated in a wet bath • Use of continuous fiber • Fiber tension assists consolidation • Ambient cure	• Quality control of mix • Wet-out • Non-uniform resin distribution • Control of cure • Environmental issues
Automated Prepreg Winding	• Automated winding of prepreg tow • Control of incoming material • Use of continuous fiber • Elevated temperature cure • Control of tension	• Prepreg cost and shelf life • Use of a machine • Space limitations
Adhesive Bonding of fabricated Shells or Plates	• Use of prefabricated sections • Adhesive bonding in the field • Rapid procedure • Ease of fabrication	• Lack of fiber continuity • Shear lag effect • Durability of the adhesive • Increase in overall thickness
In-Situ Resin Infusion	• Placement of dry fabric followed by infusion under vacuum • Fills cracks • Non-uniform geometries	• Wet-out • Excessive resin use • Appearance • Non-uniform and low tension • Difficulty of holding vacuum

Table 7 - *Comparative Analysis of FRP Reinforcement for Concrete (Adapted from Erki and Rizkalla 1998)*

Advantages	Disadvantages
a) High strength-weight ratio (10 to 15 times of that of steel) b) Excellent corrosion resistance c) Excellent electromagnetic neutrality d) Excellent fatigue characteristics (3 times of that of steel for carbon and aramid fibers) e) Low coefficient of thermal expansion	a) Higher cost b) Low modulus of elasticity for glass and aramid fibers c) Low rupture strain d) Long-term strength that can be lower than short-term static strength e) Susceptibility to ultra-violet radiation damage f) Durability of aramid fibers in water g) Unknown durability of glass fibers in alkaline environments h) Questionable fatigue characteristics for glass fibers

Tendons

The properties of FRP composites (i.e., elastic behavior, high strength and low elastic modulus) are better utilized when the reinforcement is pre- or post-tensioned (Nanni 1998). FRP tendons are reinforced with carbon (PAN and Pitch based), aramid and E-glass fibers with fiber volumes ranging from 45 to 70% (Rizkalla and Abdelrahman 1998). Different shapes exist, such as bars, cables (consisting of 7, 19 and 37 strands), rectangular strips and braided reinforcement. The tensile characteristics of FRP tendons depend on (Rizkalla and Abdelrahman 1998): a) Type and volume fraction of fiber and matrix; b) Method of production; c) Diameter of reinforcement (larger diameter bars have less tensile strength than the small diameter bars); and d) Gripping system.

CFRP and AFRP tendons are used for pre-tension and post-tension of concrete, however GFRP tendons are not recommended for pre-tension due the low resistance to alkaline environments (Sen et al. 1993). Unlike steel tendons, and due to orthotropic properties of composites, FRP tendons have considerably less lateral strength compared to axial strength. For this reason, the effect of lateral confinement and stress concentrations at the mechanical anchorage could control the allowable jacking stress (Abdelrahman et al. 1995).

Composite materials have been efficiently used as prestressed tendons in bonded (pre-tensioned) or un-bonded (post-tensioned) concrete beams. Some of the critical issues in these applications are prestress losses, composite relaxation, tensile fatigue, concrete-composite bond, transfer and development lengths, and the requirements for mechanical anchorage to apply or sustain the load. For example, in a tendon anchorage, we may expect the material to be subjected to combined stresses that require special consideration.

The lower elastic modulus of FRP tendons compared to steel tendons, results in lower prestress losses due to concrete deformation in terms of creep, shrinkage, and elastic shortening, or anchorage seating. Relaxation of FRP tendons differs greatly according to the type of fiber and matrix (ACI 1996). In general, relaxation of GFRP and CFRP tendons are comparable to that of prestressing steel, however relaxation of AFRP is much higher (Rizkalla and Abdelrahman 1998). FRP composite tendons can experience the stress-rupture or creep-rupture phenomenon, which is failure under sustained constant load. The required time to develop creep-rupture in the FRP reinforcement increases by decreasing the applied tensile stress. For the same applied tensile stress, the time for creep-rupture failure for various types of fiber reinforcement is ranked from largest to smallest as: CFRP, AFRP and GFRP (Umoto et al. 1995).

External Reinforcement of Deck Soffits and Girders

The use of techniques associated with the external attachment of composite plates to the soffit of decks and the underside of beams for purposes of strengthening and retrofit is attractive due to factors related to ease of access and decreased need for extensive changes to existing structures. Although bonding of steel plates has been used extensively for over two decades, this methods suffers from a number of disadvantages ranging from difficulty in placement, to concerns related to overall durability. Composite plates, on the other hand, do not suffer from most of these deficiencies, due to the high stiffness- and strength-to-weight ratios, corrosion resistance and lightweight.

Beginning with the repair of the Ibach Bridge in Switzerland, a number of retrofit and strengthening projects have been completed in Europe and Japan, along with a few demonstration projects in the United States. In recent years, this method has been used for the upgrading of bridge decks to enable the use of higher load levels such as in the case of Hiyosigura Viaduct in Chiba, Japan where two layers of carbon fiber unidirectional fabric were bonded to the underside of a deck section in order to increase capacity from a 20 ton level to a 25 ton level (Karbhari 1998). The use of composites in such applications provides yet another example of how lightweight tailored materials can be used effectively to enhance and increase the life of existing civil infrastructure. Details related to use and testing of such applications can be found in McKenna and Erki (1994) and Karbhari et al. (1997b). The "plate" (or external reinforcement) can itself be fabricated in three generic ways as follows:

a) *Adhesive Bonding*: Composite strip/panel is pre-manufactured and cured (using wet layup, pultrusion or autoclave cure) and then bonded onto the concrete substrate using an adhesive under pressure

b) *Wet Layup*: Resin is applied to the concrete substrate and layers of fabric are then impregnated in place using rollers and squeegees. The composite and bond are formed at the same time

c) *Resin Infusion*: Reinforcing fabric is placed at the spot to be retrofit and the entire area is encapsulated in a vacuum bag. Resin is then infused into the assembly with compaction taking place under vacuum pressure. Unlike the wet layup process this is a closed process and the infusing resin can fill cracks and voids as well.

Of these, the pre-manufactured alternative shows the highest degree of uniformity and quality control for the reinforcement strip, since it is fabricated under controlled conditions. Application is still predicated by the use of an appropriate adhesive and through the achievement of a good bond between the concrete substrate and the composite adherent. Care must be taken to ensure that the adhesive is chosen so as to match both the concrete and the composite and provide an interlayer to reduce mismatch-induced stresses. Bonding can be assisted in this case through the temporary use of external clamps or a vacuum bag to provide compaction pressure.

The wet layup process is perhaps the most used and gives the maximum flexibility for field application, and is probably also the least expensive alternative. However, it presents the largest variability in properties and necessitates the use of excessive resin, and could result in wrinkling of the fabric used, as well as entrapment of air.

The in-situ infusion method is a fairly new variant and is capable of achieving uniformity and good fabric compaction, while making it easier for the reinforcement to be made to fit the exact contours of the component to be strengthened. In the latter two systems the resin takes the adhesive function itself with the bond to the concrete substrate being formed simultaneously with the processing of the composite. This is both an advantage and a disadvantage, since the elimination of a third phase, the adhesive, results in fewer interfaces at which failure could occur, but also eliminates the use of a more compliant interlayer.

Retrofit of Walls

The seismic repair and retrofitting of reinforced and unreinforced concrete and masonry walls has been shown to be possible through the use of thin composite overlays (Seible 1995, Priestly and Seible 1995, Ehsani et al. 1997) with fibers oriented horizontally. Tests have shown that the application of one to two layers of unidirectional fabric can result in significant enhancements in performance with reductions in shear deformation, and increases in ductile flexural in-plane behavior. The application of this technology to deficient wall systems, and to systems where cracking is due to uneven settlements is also possible.

Seismic Retrofit of Bridge Columns

The confinement of concrete through the use of an external jacket has been proven to result in enhanced system ductility and subsequent performance under seismic events. This is because the external jacket provides constraint to the dilation of concrete in the absence of sufficient hoop steel. Amongst other conventional seismic retrofit strategies, steel jackets have been extensively used in both Japan and the United States. However, this method of retrofit necessitates the welding of jacket sections in the field, which is a concern as related to effectiveness and overall quality control. Further the use of steel in some areas carries with it the potential for rapid corrosion. The use of composites as wraps/jackets on columns to replace steel jackets in seismic retrofit has achieved a high level of interest both in Japan and the United States, with the generic methods used to achieve the confinement of concrete including: a) Wet winding of tows; b) Winding of prepreg tow/tape; c) Wet layup of fabric; d) Layup of tape; e) Adhesive bonding of prefabricated shells; f) In situ resin infusion of jackets; and g) Use of composite cables wrapped around a concrete core.

A discussion on the development of design guidelines, as well as a comparison of the different methods based on materials choice and method of fabrication is given in Seible and Karbhari (1997). As an example of development, tests on 40% scale bridge columns at the University of California, San Diego (UCSD) have shown that the use of carbon fiber reinforced jackets applied through the continuous winding of prepreg tow which is cured after completion of the jacket, can be as effective as steel jackets for retrofits in flexural lap splice and shear applications (Seible et al. 1997). The advantages of such a method are: a) The fiber is continuously placed in the direction where it is the most efficient, i.e., the hoop direction; b) The method of application lends itself to overall efficiency with significantly lower thickness of jacketing required; and c) The use of an automated method of construction leads to a high degree of quality control at a rapid rate of completion. The use of composites, especially as reinforced with carbon fibers in the hoop direction, provides a layup that is tailored for the specifics of the application, resulting in materials efficiencies that were not possible with conventional materials. The entire jacket is coated with a layer that served both as protection and for aesthetic. Extensive testing has been conducted using this technology with both 40% scale and full-scale validation tests. In addition, successful field demonstration applications have been conducted on the I-10 Santa Monica Viaduct in Los Angeles and a comprehensive set of design guidelines have been developed (Seible et al. 1996).

The use of composites to retrofit concrete columns has been shown to be cost-competitive with steel retrofits while providing the potential for faster fabrication with less lane closure and potentially far greater durability over the lifetime of use. Further details related to testing and materials selection are reported in (Karbhari 1997). For purposes of illustration of materials efficiency, the relative thickness of jackets fabricated using E-glass/epoxy, S-glass/epoxy, Graphite/epoxy, Kevlar/epoxy and Boron/epoxy at 60% fiber volume fraction are shown in Table 8 as a function of the retrofit mechanism.

Table 8- *Comparison of Hypothetical Normalized FRP Jacket Thicknesses for Column Retrofit*

Material	Properties[a] for $V_f = 60\%$	Normalized Jacket Thickness			
		Shear Strengthening	Plastic Hinge Confinement	Lap Splice Clamping	Bar Buckling Restraint
Proportionality Relationship		$t_j^v \sim \frac{1}{E_j D} \bullet C_v$	$t_j^c \sim \frac{D}{f_{ju} \varepsilon_{ju}} \bullet C_c$	$t_j^b \sim \frac{D}{E_j} \bullet C_b$	$t_j^s \sim \frac{D}{E_j} \bullet C_s$
E-glass /epoxy	E = 45 GPa σ =1020 MPa ε= 2.3%	1	1	1	1
S-glass /epoxy	E = 55 GPa σ =1620 MPa ε= 2.9%	0.82	0.50	0.82	0.82
Kevlar 49 /epoxy	E = 76 GPa σ =1380 MPa ε= 1.6%	0.59	1.06	0.59	0.59
Carbon /epoxy	E = 160 GPa σ =1725 MPa ε= 0.9 %	0.28	1.51	0.28	0.28
Boron /epoxy	E = 210 GPa σ =1240 MPa ε= 0.6 %	0.21	3.15	0.21	0.21

[a]Values are representative averages without application of reduction coefficients for aging and environmental durability

Replacement Bridge Deck Systems

Of all elements in a bridge superstructure, bridge decks may perhaps require the maximum maintenance, for reasons ranging from the deterioration of the wearing surface to the degradation of the deck system itself. Added to the problems of deterioration are the issues related to the need for higher load ratings (HS20 to HS25, for example) and increased number of lanes to accommodate the ever-increasing traffic flow on major arteries. Beyond the costs and visible consequences associated with continuous retrofit and repair of such structural components are the real consequences related to losses in productivity and overall economies related to time and resources caused by delays and detours (See for example Ehelen and Marshall 1996). Reasons such as those listed above provide significant impetus for the development of new bridge decks out of materials that are durable, light and easy to install. Besides the potentially lower overall life-cycle costs (due to decreased maintenance requirements), decks fabricated from fiber reinforced composites are significantly lighter, thereby affecting savings in substructure costs, enabling the use of higher live load levels in the case of replacement decks, and bringing forth the potential of longer unsupported spans and enhanced seismic resistance.

FRP composites have been used for decks, beams and superstructures of pedestrian and highway bridges. Two of the first FRP composite bridges were the Miyun Bridge built in China in 1982 (Mufti et al. 1991), and the Aberfeldy cable-stayed footbridge built in Scotland in 1992 (FHWA 1997). In the US, Lockheed Martin designed and fabricated a 9.1-m FRP bridge superstructure, which was subjected to extensive testing (Rodriguez et al. 1997). Most recently, the focus on FRP composite bridge developments focused on decks supported by concrete and steel girders (Zureick et al. 1995, Karbhari 1996, Lopez-Anido and GangaRao 1997, Salim et al. 1997, Soneji et al. 1999). Several FRP deck systems have been developed and used in demonstration projects both for replacement of aged concrete or wood decks, as well as for new construction.

An evaluation plan for FRP bridge decks has been recently proposed by the HITEC program (Seible et al. 2000). The panel has identified key technical issues and proposed performance verification tests related to: a) Structural system response, inspection, maintenance and repair; b) Joints and connections; and c) Materials and manufacturing. Currently, the Ohio Department of Transportation is conducting an evaluation program for four different FRP bridge deck systems, which ranges from durability characterization and structural fatigue testing to field installation and monitoring (Triandafilou 2000). In general FRP bridge decks can be classified in three groups:

a) FRP decks fabricated by interlocking pultruded profiles – In this type of decks, components with constant cross-section are fabricated by the pultrusion process. In a second operation the components are interlocked and bonded using a toughened adhesive. In this way large panels can be fabricated and shipped to the construction site. This type of decks is typically highly orthotropic with the main stiffness direction corresponding to the axis of the pultruded profiles (Lopez-Anido and GangaRao 1997, Harik et al. 1999a).

b) FRP decks fabricated based on the concept of sandwich construction – In this case, the deck is formed by bonding stiff and strong face sheets on top and bottom of a shear-resistant core. Typically the core has vertical cells with different configurations and filler materials (foam, balsa wood). At least two fabrication processes have been applied to fabricate sandwich type decks: VARTM and hand layup compaction molding (Harik et al. 1999c, 2000).

c) FRP-concrete hybrid decks – This is a hybrid deck systems that uses FRP panels not only for tension reinforcement of concrete, but also as a stay-in-place form (Lopez-Anido et al., 1999, Harik et al. 1999b).

Typically fiber-composite bridge decks provide almost linear elastic response to failure, with load levels significantly higher than those achieved by reinforced concrete deck panels, but with comparable stiffnesses, at weights one-fourth of those of reinforced concrete incumbents. Therefore, the design of FRP bridge decks is typically stiffness driven, and controlled by deflection limits. Examples of FRP composite deck systems that have been used for bridge construction are presented in Table 9.

Table 9 - *Examples of Short-Span Fiber-Composite Bridge Decks Installed*

Name / Location	Supplier / State	Opened	FRP Deck System	Supporting Beams
Laurel Lick Bridge, WV	Creative Pultrusions, PA	May 1997	Interlocking pultruded components	FRP composite pultruded beam
Magazine Ditch Bridge, DE	Hardcore DuPont, DE	July 1997	Vertical cells by VARTM	Edge pre-stress concrete beams
Tech 21, Butler County, OH	Martin Marietta Materials, NC	July 1997	Pultruded and lay-up sections	FRP composite U-shaped beams
Wickwire Run Bridge, WV	Creative Pultrusions, PA	Sept. 1997	Interlocking pultruded components	Steel wide flange girder

New Structural Systems: Composite Shells

Although the areas of rehabilitation and retrofit offer the maximum potential for immediate application of composites in civil infrastructure, the development of new systems that combine the directionality and high performance levels of composites with the dominant characteristics of conventional materials (such as with concrete in compression) show great potential for advances in the design and construction of new civil engineering structures. One such concept is that of composite shell systems for columns, wherein prefabricated composite tubes serve the dual purpose of formwork and reinforcement, thereby replacing the reinforcing steel while enabling faster construction. In this system, the column is constructed by placing the hollow composite shell in place, and then filling it with concrete. Construction details are such that it can be directly incorporated with conventional construction methods. The composite shell system has been successfully tested using two concepts: a) Wherein starter bars from the footing extended into the column providing a conventional anchorage mechanism into the footing and/or bridge superstructure; and b) Wherein steel is completely eliminated with anchorage force mechanism consisting of the reaction moment due to the compression force couple generated inside the footing and the frictional stresses between the composite shell and concrete (Seible et al. 1996).

Durability

Composites, like other man-made materials, degrade with time. The goal is to engineer the material to extend the service life of infrastructure applications. This proposition requires a comprehensive evaluation of aging rates. One of the main drawbacks to the use of composites in infrastructure has been the lack of data on performance and durability. Building a database on structural behavior and aging of composites is the first step to gain acceptance of the material. Durability characterization of composites has to be done at micro level, coupon level, component level, and system level. These evaluations have to be carried out by systematically varying the temperature, ranging from -40°F (-40°C) to 160°F (71°C); pH value, ranging from 3 to 13; relative humidity, ranging from 20% to 98%, and at varying pressures depending upon materials and their use requirements (GangaRao and Lopez-Anido 1996).

The importance of micro level testing under the varying parameters stated above is such that the results from dynamic mechanical and thermal analyzing, differential calorimetric scanning, fractography through scanning electronic microscopy and through other methods would reveal the initiation of degradation either of the matrix or the fiber or the interface. In this way, the early detection of degradation contributes to the understanding of aging rates of composite materials in civil infrastructure.

Current Status and Future Research Needs

Two problems that the authors have been confronted with when using composites in infrastructure projects are the repeatability of properties of mass-produced composite products and the compliance with construction tolerances. Coordinated efforts are required to achieve significant advancements in the development of composite material specifications and quality-control guides. The establishment of a program to qualify specific construction and repair composite materials is essential to expedite the use of composites in infrastructure.

Infrastructure construction and repair with composites and hybrids may very well be a cost-effective alternative to conventional technologies. However, it is very unlikely that their use will be accepted unless techniques are developed for testing and evaluating their in-service performance to insure continued integrity. Recent advances in ultrasonics, acoustic emission and infrared thermography testing and data interpretation techniques have opened up potential application opportunities in the area of defect detection in structural composites (for example, cracks at the bond-line between composite fabrics and concrete members may be spotted at an early stage).

The challenge facing composites in infrastructure is to straddle two technologies: civil engineering practice and composite materials design and fabrication. This challenge is reflected by the current status on structural design with FRP composites (Ellingwood 2000), as shown in Table 10.

Table 10 – *Current Status of Structural Design with FRP Composites (Ellingwood 2000)*

Issue	**Status**
Material availability	broad
Product diversity	wide
Level of technology	high
Cost	becoming competitive
Structural mechanics	reasonably well understood
Understanding of behavior	uneven
Supporting databases	limited
Standards and specifications	lacking
Structural design methodology	primitive
Confidence on reliability of structural product	marginal

To address these issues, government agencies, international groups and industry have identified key research needs. Scalzi et al. (1999) emphasized the need to develop material/product standards and performance criteria. Bank (1997) indicated the need for coordinating research activities at the international level to avoid duplication and build on each other's progress. The IRACC Workshop Report (Uomoto et al. 1997) stressed the need for continuous monitoring, durability data, education, codes and databases, among others.

The NIST-ATP program on Composites in Civil Applications (Wu 1998) identified specific issues to be addressed: 1) The development of continuous composite fabrication processes, from fiber placement through resin impregnation and cure; 2) Automated methods of component assembly into large scale structures; 3) The impact of automation and production volume on raw material and production costs; 4) Methodologies for design and utilization of pre-manufactured structural elements assembled at the worksite instead of the build in place model used today; and 5) Methods to build in health monitoring systems for subsequent field inspection.

Summary and Conclusions

The evaluation of composites in infrastructure is a three-fold task that encompasses the elaboration of design guidelines and specifications, the understanding of the long-term performance, and the development of construction and repair technologies. In the near future we may expect a proliferation of composite material applications in several infrastructure areas. Among the most promising applications, we shall mention bridge decks and rehabilitation, retrofit of concrete structures and engineered wood products. The challenge for composites is to out-perform conventional construction materials and technologies and at the same time be cost-competitive.

The extent of composites applications will depend in large part on the resolution of outstanding critical issues that include: a) Durability and fire resistance; b) Reparability of composite structural elements; c) Development of validated specifications, standards, and guidelines of use to civil engineers; d) Development of practical design methods and cost-effective manufacturing processes that optimize the use of the material; and e) Provision of an appropriate level of quality assurance and control both during manufacturing and installation by contractors.

Acknowledgements

The authors are thankful to several colleagues researching topics related to composites in infrastructure at the University of California San Diego, West Virginia University and the University of Maine for many valuable discussions that contributed to formalize the concepts presented in this paper. The authors also wish to express their gratitude to DOT engineers from California, West Virginia, Ohio and Maine for their interest in pursuing innovative composite demonstration project and developing practical installation methods.

References

Handbooks, Guidelines and Specifications for Composites in Civil Infrastructure

American Concrete Institute (ACI), (1998a), *Provisional Design Recommendations for Concrete Reinforced with FRP Bars*, Committee 440, Farmington Hills, MI.

American Concrete Institute (ACI), (1998b), *Guidelines for Selection, Design, and Installation of Fiber Reinforced Polymer Systems for Externally Strengthening Concrete Structures,* Committee 440, Farmington Hills, MI.

American Concrete Institute (ACI), (1998c), *Draft Standard Test Methods for FRP Rod and Sheet*, Committee 440, Farmington Hills, MI.

American Society of Civil Engineers (ASCE), (1984), *Structural Plastics Design Manual*, Task Committee on Design, Structural Plastics Research Council, New York, NY.

American Society for Testing and Materials (ASTM), (2000), *Committee D-30 Composite Materials*, West Conshohocken, Pennsylvania.

Canadian Standards Association (CSA), (1996), *Canadian Highways Bridge Design Code, Section 16, Fiber Reinforced Structures (CHBDC)*, Toronto, Canada, 28 pp.

Clarke, J.L. (ed.), (1997), *EUROCOMP Design Code and Handbook, Structural Design of Polymer Composites*, E&F Spon, London, UK.

Composite Materials Handbook-MIL-17, (1999), *Polymer Matrix Composites Handbook*, MIL-HDBK-17 PMC, Vol. 1-3, Army Research Laboratory, Aberdeen Proving Ground, MD.

Duthinh, D., (1999), *NIST Workshop on Standards Development for the Use of Fiber Reinforced Polymers for the Rehabilitation of Concrete and Masonry Structures, Proceedings, January 1998, Tucson, Arizona*, Publication No. NISTIR 6288, National Institute of Standards and Technology, Gaithersburg, MD.

Ellingwood, B.R. (2000), *Load and Resistance Factor Design (LRFD) for Structures using Fiber-Reinforced Polymer (FRP) Composites*, (Presentation at ICE'99, May 1999, Cincinnati, OH), NIST Publication No. GCR 00-793, National Institute of Standards and Technology, Gaithersburg, MD.

Hart-Smith, L. J., (1982), *Design Methodology for Bonded-Bolted Composite Joints*, AFWAL-TR-81-3154, Vol. 1 and 2, Air Force Flight Dynamics Laboratory, Wright-Patterson, AFB, OH.

International Conference of Building Officials (ICBO), (1997), *Acceptance Criteria for Concrete and Reinforced And Unreinforced Masonry Strengthening Using Fiber-Reinforced, Composite Systems*, AC125, ICBO Evaluation Services, Whittier, CA.

Japan Society of Civil Engineers (JSCE), (1997), *Recommendation for Design and Construction of Concrete Structures Using Continuous Fiber Reinforcing Materials,* Concrete Engineering Series, No. 23, Tokyo, Japan, 325 pp.

Ramkumar, R.L., Saether, E.S. and Cheng, D., (1986), *Design Guide for Bolted Joints in Composite Structures*, AFWAL-TR-86-3035, Air Force Flight Dynamics Laboratory, Wright-Patterson AFB, Ohio.

Suppliers of Advanced Composite Materials Association (SACMA), (1996) *Safe Handling of Advanced Composites*, 3rd Edition, Arlington, VA.

Overview and Demonstration Projects of Composites in Civil Infrastructure

Abdelrahman, A.A., Tadros, G., and Rizkalla, S.H., (1995), "Test Model for the First Canadian Smart Highway Bridge," ACI Structural Journal, V. 92, No. 4.

American Concrete Institute (ACI), (1996). State-of-the-Art Report on Fiber Reinforced Plastic (FRP) Reinforcement for Concrete Structures, Report ACI 440R-96, Farmington Hills, MI.

Ballinger, C.A., (1997), "Infrastructure Repair and Retrofit Strengthening with External FRP Systems," *Composites Fabrication*, October, pp. 18-22.

Bassett S. (ed.) (1998), *Composite for Infrastructure – A Guide for Civil Engineers*, Ray Publishing, Wheat Ridge, CO.

BIRL / Northwestern University BIRL, (1995), *Definition of Infrastructure Specific Markets For Composite Materials : Topical Report*, Project P93-121 / A573, Evanston, IL.

Busel, J. (ed.), (1995), *FRP Composites in Construction Applications – A Profile in Progress,* Market Development Alliance of the Composites Institute, New York, NY.

Civil Engineering Research Foundation (CERF), (1994), *Materials for Tomorrow's Infrastructure: Technical Report: A Ten Year Plan for Deploying High Performance Construction Materials and Systems*, CERF Report No. 94-5011E, Washington, DC.

Erki, M.A. and Rizkalla, S.H., (1998), "FRP Reinforcement for Concrete Structures," *ACI Compilation: Fiber Reinforced Plastic Reinforcement*, American Concrete Institute, Farmington Hills, MI, pp. 3-8.

Fletcher, A., (1994), Advanced Composites - A Profile of the International Advanced Composites Industry, Elsevier Science, New York, NY.

GangaRao, H.V.S. and Lopez-Anido, R. (1996). *Evaluating Composites For Infrastructure Applications,* CDA (Composites Design & Application), published by the Composites Institute of the Society of Plastics Industry, Summer Issue, pp. 24-31.

GangaRao, H.V.S., Thippeswamy, H.K., Kumar, S.V., and Franco, J.M., (1997), "Design, Construction and Monitoring of the First FRP Reinforced Concrete Bridge Deck in the United States," *Proceedings of the Third International Symposium on Non-Metallic* (FRP) Reinforcement for Concrete Structures (FRPRCS-3), Sapporo, Japan, Vol. 1, pp. 647-656.

Scalzi J.B., Podolny, W., Munley, E. and Tang, B., (1999), *Will FRP Composites Ever Change Bridge Engineering*?, Joint National Science Foundation-Federal Highway Administration Report, Washington, DC.

Wu, H.F. (1998), Composites in Civil Applications, Executive Summary, National Institute of Standards and Technology, Gaithersburg, MD.

International Perspective on Composites in Civil Infrastructure

Bank, L.C., (1997), *Project Technical Report on the US Delegation to the Third International Symposium on Non-Metallic (FRP) Reinforcement for Concrete Structures (Sapporo, Japan)*, National Science Foundation, Arlington, VA.

Federal Highway Administration (1997), *FHWA Study Tour for Advanced Composites in Bridges in Europe and Japan*, U.S. Department of Transportation, Washington, DC.

Karbhari, V.M., (1998), *Use of Composite Materials in Civil Infrastructure in Japan*, World Technology Monograph, International Technology Research Institute, Baltimore, MD.

Mufti, A. A., Erki, M-A., and Jaeger, L. G., (1991), *Advanced Composite Materials with Application to Bridges*, State of the Art Report, The Canadian Society for Civil Engineers, Montreal, Canada.

Nanni, A., Di Tommaso, A., Arduini, M. and Cheng, J.J.R. (1996), *International Research on Advanced Composites in Construction*, Report No. IRACC-96, NSF, Arlington, VA.

Nanni, A. (1998), *RC Structures with FRP Materials,* Lecture Notes for ICCI Short Course on Design and Retrofit of Structures with Fiber Composites, Tucson, AZ.

Rizkalla, S. and Abdelrahman, A. (1998), *Design of Concrete Members Prestressed by FRP,* Lecture Notes for ICCI Short Course on Design and Retrofit of Structures with Fiber Composites, Tucson, AZ.

Seible, F., Hegemier, G.A., Karbhari, V.M., Davol, A., Burgueno, R., Wernli, M. and Zhao, L., (1996), *The I-5/Gilman Advanced Composite Cable Stayed Bridge Study*, University of California, San Diego, Report No. SSRP-96/05, San Diego, CA.

Taerwe, L., (1997), "FRP Activities in Europe: Survey of Research and Applications," *Proceeding of the Third International Symposium on Non-Metallic (FRP) Reinforcement for Concrete Structures (FRPRCS-3),* Sapporo, Japan, Vol. 1, pp. 59-74.

Uomoto, T., Nanni, A., Di Tommaso, A., Fukuyama, H., and Arduini, M., (1997), *Minutes of the Fourth International Research on Advanced Composites in Construction Workshop*, IRACC-97, NSF, Arlington, VA.

Wilkins, D.J. (ed.), (1994), *JTEC Panel Report on Advanced Manufacturing Technology for Polymer Composite Structures in Japan*, Japanese Technology Evaluation Center, Baltimore, MD, NTIS PB94-161403.

FRP bridge decks

Banuk, R., (1996), "Aircraft Company's Venture into Non-Traditional Composite Structure: Bridge Decks, Windmills, Pedestrian Bridge, Transit Bus." American Society of Mechanical Engineers, Aerospace Division (Publication), Proceedings of *1996 ASME International Mechanical Engineering Congress and Exposition*, Vol. 52, Atlanta, GA, p. 89.

Ehelen, M. A., and Marshall, H. E., (1996). *The Economics of New-Technology Materials in Construction: A Case Study of FRP Composite Bridge Decking*, Report NISTIR 5864, National Institute of Standards and Technology, Gaithersburg, MD.

Harik, I., Alagusundaramoorthy, P., Siddiqui, R., Lopez-Anido, R., Morton, S., Dutta, P. and Shahrooz, B., (1999a), "Static Testing on FRP Bridge Deck Panels," in *Evolving and Revolutionary Technologies for the New Millennium*, Society for the Advancement of Material and Process Engineering, Vol. 44, Covina, CA, pp. 1643-1654.

Harik I., Alagusundaramoorthy P., Siddiqui R., Lopez-Anido R., Morton S., Dutta P., Shahrooz B., (1999b), "Testing of Concrete/FRP Composite Deck Panels," in *Materials and Construction: Exploring the Connection*, Ed. L. Bank, ASCE, Reston, VA, pp. 351-358.

Harik, P. Alagusundaramoorthy, Siddiqui, R. Lopez-Anido, R. Morton, S., Dutta, P. and Shahrooz, B., (1999c), "Testing of Fiberglass Composite Bridge Deck Panels," *Second International Conference CMT 99-Computational Methods and Testing for Engineering Integrity*, Eds. Chen, C.S., Brebbia, C.A., and Pepper, D.W., WIT Press, Southampton, UK, pp. 663-672.

Harik, P. Alagusundaramoorthy, Siddiqui, R. Lopez-Anido, R. Morton, S., Dutta, P. and Shahrooz, B., (2000), "Testing of Contact Molding Hand Lay-Up FRP Deck Panels," Innovative Process and Material Solutions for Creative Design Effectiveness, Ed. M.E. Erath, SAMPE Europe, Aalsmeer, Netherlands, pp. 425-434.

Howdyshell, P.A., Trovillion, J.C., GangaRao, H.V.S. and Lopez-Anido, R., (1998), *Development and Demonstration of Advanced Design Composite Structural Components*, U.S. Army Corps of Engineers, Construction Engineering Research Lab., Technical Report CERL-TR-98/99, Champaign, IL.

Karbhari, V.M., (1996), "Fiber Reinforced Composite Decks for Infrastructure Renewal," *Advanced Composite Materials in Bridges and Structures*, Montreal, Canada, pp. 759-766.

Karbhari, V.M. (1997) "On the Use of Composites for Bridge Renewal," Materials, Manufacturing and Durability, *Proceedings of the 42nd. International SAMPE Symposium*, pp. 915-926.

Karbhari, V.M., Seible, F., Hegemier, G.A., and Zhao, L., (1997a) "Fiber Reinforced Composite Decks for Infrastructure Renewal – Results and Issues," *Proceedings of the International Composites EXPO '97*, Session 3-C, pp. 1-6.

Lopez-Anido, R. and GangaRao H.V.S., (1997), "Design and Construction of Composite Material Bridges," in *Advanced Rehabilitation, Durable Materials, Non-Destructive Evaluation and Management*, Eds. U. Meier & R. Betti, Zurich, CH, pp. 269-276.

Lopez-Anido, R., Stephenson, L. D. and Howdyshell, P. (1998), "Durability of Modular FRP Composite Bridge Decks Under Cyclic Loading," in *Durability of FRP Composites for Construction*, Eds. B. Benmokrane and H. Rahman, Sherbrooke, Canada, pp. 611-622.

Lopez-Anido, R., Dutta, P., Bouzoun, J., Morton, S., Shahrooz and B., Harik I., (1999), "Fatigue Evaluation of FRP-Concrete Bridge Deck on Steel Girders at High Temperature," in *Evolving and Revolutionary Technologies for the New Millennium*, Society for the Advancement of Material and Process Engineering, Vol. 44, Covina, CA, pp. 1666-1675.

Mertz, D.R. and Kulicki J.M. (1997), "The Application of the AASHTO LRFD Bridge Design Specifications to Composite-Material Bridges," *in Advanced Rehabilitation, Durable Materials, Non-Destructive Evaluation and Management*, Eds. U. Meier & R. Betti, Zurich, CH, pp. 178-184.

Rodriquez, J. G., Dumlao, C., Coilko, A. T., Pfister, T.J., Schmeckpeper, E. R., (1997) *Test and Evaluation Plan for the Composite Bridge.* Idaho National Engineering Laboratory Transportation Infrastructure Department Lockheed Martin Idaho Technologies Company, Idaho Falls, ID.

Salim, H.A., Davalos, J.F., Qiao, P., Kiger, S., (1997) "Analysis And Design Of Fiber Reinforced Plastic Composite Deck-And-Stringer Bridges," Proceedings of the 9th International Conference on Composite Structures, ICCS/9, Elsevier Science Ltd., England, Vol. 38, No. 1-4, pp. 295-307.

Seible, F., R. Christie, C. Hiel, J. Hooks, V. Karbhari, M. Karshenas, P. Liles, R. Lopez-Anido, S. Morton, T. Post, C. Schutte, C. Seim, R. Till, D. Williams and A. Yannotti, (2000) FRP Bridge Decks Evaluation Panel, *FRP Composite Bridge Decks*, HITEC Evaluation Plan Civil, Engineering Research Foundation, Washington, DC.

Soneji, J., C. Hu, M. Chaudhri, A. Faqiri, J. W. Gillespie Jr., D. A. Eckel II, D. R. Mertz, and M. J. Chajes, (1999), "Use of Glass-Fiber-Reinforced Composite Panels to Replace the Superstructure for Bridge 351 on N387A Over Muddy Run," 16th Annual International Bridge Conference, Pittsburgh, PA.

Triandafilou, L., (2000), "The Great Miami FRP Composite Bridge Deck Project", Report by the Federal Highway Administration - Eastern Resource Center, Presented at the 5th World Pultrusion Conference, European Pultrusion Technology Association, Berlin.

Zureick, A. H., Shih, B. and Munley, E., (1995). "Fiber-Reinforced Polymeric Bridge Decks," *Structural Engineering Review*, Vol. 7, No. 3, Pergamon Press Ltd., pp. 257-266.

Structural Rehabilitation and Seismic Retrofit

Ehsani, M.R., Saadatmanesh, H., Al-Saidy, A., (1997), "Shear Behavior of URM Retrofitted with FRP Overlays," *Journal of Composites for Construction,* Vol. 1., No.1, pp.17-25.

Gergely, J., Pantelides, C.P., and Reaveley, L.D., (2000), "Shear Strengthening of RCT-Joints Using CFRP Composites," *Journal of Composites for Construction*, ASCE, Vol. 4, No. 2, pp 56-64.

McKenna, J.K. and Erki M.A. (1994). "Strengthening of Reinforced Concrete Flexural Members Using Externally Applied Steel Plates and Fiber Composite Sheets - A Survey," *Canadian Society of Civil Engineers*, Vol. 21, No.1, pp. 16-24.

Priestly, M.J.N and Seible, F. (1995). "Design of Seismic Retrofit Measures for Concrete and Masonry Structures," *Construction and Building Research*, Vol. 9, No. 6, pp. 365-377.

Saadatmanesh, H., Ehsani, M.R. and Li, M.W., (1994), "Strength and Ductility of Concrete Columns Externally Reinforced with Fiber Composite Straps," *ACI Structural Journal*, Vol. 91, No. 4, pp. 434-447.

Seible, F., (1995), "Structural Rehabilitation with Advanced Composites," *Proceedings of the IABSE Symposium*, San Francisco, CA, pp. 391-398.

Seible, F. and Karbhari, V.M., (1997), "Seismic Retrofit of Bridge Columns using Advanced Composite Materials," *Proceedings of the National Seminar on Advanced Composite Material Bridges*, Arlington, VA, 29 pp.

Seible, F., Priestley, M.J.N, Hegemier, G., and Innamorato, D., (1997), "Seismic Retrofit of RC Columns with Continuous Carbon Fiber Jackets," *Journal of Composites in Construction,* ASCE, Vol. 1, No. 2, pp. 52–62.

Connections

Bank, L.C., Mosallam, A.S., and McCoy, G.T., (1993), "Design and Performance of Connections for Pultruded Frame Structures," *Journal of Reinforced Plastics and Composites*, 13(3), pp. 199-212.

Bank, L.C., Yin, J., Moore, L.E., Evans, D. and Allison R., (1996), "Experimental and Numerical Evaluation of Beam-to-Column Connections for Pultruded Structures," *J. Reinforced Plastics & Comp.*, Vol.15, No.10, pp.1052-1067.

Mottram, J.T., and Zheng, Y. (1996), "State-of-the-art review on the design of beam-to-column connections for pultruded frames," *Composite Structures*, Vol. 35, Issue 4, pp. 387-401.

Turvey, G.J., (1997), "Analysis of pultruded GFRP Beams with Semi-Rigid end Connections," *Composite Structures*, Vol. 38, Iss. 1-1, pp. 3-16.

FRP Reinforced Wood

Dagher, H.J., Kimball, T., Abdel-Magid, B., and Shaler, S., (1996), "Effect of FRP Reinforcement on Low-Grade Eastern Hemlock Glulams." Proceedings: *1996 National Conference on Wood Transportation Structures*, Madison, WI.

Dagher, H.J., Poulin, J., Abdel-Magid, B., Shaler, S.M., Tjoelker, W, Yeh, B., (1998), "FRP Reinforcement of Douglas Fir and Western Hemlock Glulam Beams," Proceedings: *International Composites Expo '98*, Nashville, TN, 22C, pp.1-4.

Dagher, H.J. and Lindyberg, R., (1999), FRP Reinforced Wood in Bridge Applications. Proceedings: *1st RILEM Symposium on Timber Engineering*, Sweden, pp. 591-598.

Davalos, J.F., Barbero, E., and Munipalle, U., (1992), "Glued-laminated timber beams reinforced with E-glass/polyester pultruded composites," *Proceedings of the Structures Congress X*, ASCE, San Antonio, TX, pp.47-50.

Davalos, J.F., GangaRao, H.V.S., Sonti, S.S., Moody, R.C. and Hernandez, R., (1994), "Bulb-T and Glulam-FRP Beams for Timber Bridges." *Proceedings of the Structures Congress XII*, ASCE, Atlanta, GA, pp. 1316-1321.

Gardner, D. J., Davalos, J. F., and Munipalle, U. M., (1994), "Adhesive bonding of pultruded fiber-reinforced plastic to wood," *Forest Products J*ournal, Vol. 44, No.5, pp. 62-66.

Plevris, N. and Triantafillou, T.C. (1992), "FRP-Reinforced Wood as a Structural Material," Journal of Materials in Civil Engineering, Vol.4, No.3, pp.300-317.

Sonti, S.S. and GangaRao, H.V.S., (1996), "Banding Timber Crossties using Composite Fabrics for improving their performance," *Materials for the New Technology*, Ed. K.P. Chong, ASCE, New York, NY, Vol. 2. pp. 1449-1457.

Tingley, D.A., Gai, C. and Giltner, E. (1997). "Testing Methods to Determine Properties of Fiber Reinforced Plastic Panels used for Reinforcing Glulams," J. Composites for Construction, ASCE, pp. 160-167.

Durability

Chin, J.W., Nguyen, T., Aouadi, K., (1997), "Effects of Environmental Exposure on Fiber-Reinforced Plastic (FRP) Materials Used in Construction," *Journal of Composites Technology & Research,* American Society for Testing and Materials, pp. 205-213.

Crasto, A. S. and Kim, R. Y., (1996), "Environmental Durability of a Composite-to-Composite Adhesive Bond in Infrastructure Applications," *Proceedings of the 1996 28th International SAMPE Technical Conference*, Vol. 28, pp. 837-849, Covina, CA.

Gentry, T.R., Bank, L.C., Barkatt, A., and Prian, L., (1998), "Accelerated Test Methods to Determine the Long Term Behavior of Composite Highway Structures Subject to Environmental Loading," *Journal of Composites Technology & Research,* American Society for Testing & Materials, pp.38-50.

Karbhari, V.M., Engineer, M. and Eckel, D.A., (1997b), "On the Durability of Composite Rehabilitation Schemes for Concrete: Use of a Peel Test, *Journal of Material Science*, Vol. 32, pp. 147-156.

Karbhari, V.M., Zhao, L., Murphy, K. and Kabalnova, L., (1998), "Environmental Durability of Glass Fiber Reinforced Composites - Short Term Effects," in *Durability of FRP Composites for Construction*, Eds. B. Benmokrane and H. Rahman, Sherbrooke, Canada, pp. 513-523.

Karbhari, V.M., Chin, J.W. and Reynaud D., (2000), *Durability of FRP Composites in Civil Infrastructure*, Presented at the Transportation Research Board Meeting, NIST-CERF Program, Washington, DC.

Lesko, J. J., Hayes, M. D., Garcia, K., McBagonluri, D., Verghese, N., (1997), "Environmental –Mechanical Durability of Glass/Vinyl Ester Composites." 3rd International Conference on Progress in Durability Analysis of Composite Systems, Tucson, AZ. pp. 4.10-13.

Liao, K., Schultheisz, C.R., Hunston, D.L., Brinson L.C. (1998), "Long-Term Durabililty of Fiber-Reinforced Polymer-Matrix Composite Materials for Infrastructure Applications: A Review," *Journal of Advanced Materials*, Vol. 30, No. 4, pp. 3-40.

Umoto, T, Nishimura, T., and Ohga, H., (1995), "Static and Fatigue Strength of FRP Rods for Concrete Reinforcement," *Non-Metallic (FRP) Reinforcement for Concrete Structures*, Proceedings of the Second International RILEM Symposium (FRPRCS-2), E&FN Spon, Ghent, Belgium, pp. 100-107.

Springer, G.S., (1988), "Chapter 1 - Environmental Effects," *in Environmental Effects on Composite Materials*, Ed. Springer, G.S., Vol. 3, Technomic Publishing Co., Lancaster, PA.

Sen. R., Mariscal, D. and Shahawy, M., (1993), " Durability of Fiber Glass Pretensioned Beams," ACI Structural Journal, V. 90, No.5, pp. 525-533.

Talreja, R., (1987), *Fatigue of Composite Materials*, Technomic Publishing Co., Lancaster, PA.

Zweben, C., (1994), *Is there a Size Effect in Composites?*, Composites, Vol. 25, No. 6, pp. 451-454.

Practical Books and References on Composites

Barbero, E.J., (1998), *Introduction to Composite Materials Design,* Taylor & Francis, Philadelphia, PA.

Daniel, I.M. and Ishai, O., (1994), *Engineering Mechanics of Composite Materials*, Oxford University Press, New York, NY.

Jones. R.M., (1999), Mechanics of Composite Materials, 2nd. Edition, Taylor & Francis, Philadelphia, PA.

Mallick, P.K., (1994), *Fiber-Reinforced Composites, Materials, Manufacturing and Design*, 2nd. Edition, M. Dekker. New York, NY.

Reddy, J.N. and Miravete A. (1995) *Practical Analysis of Composite Laminates*, CRC Press, Boca Raton, FL.

Tsai, S.W., (1992), Theory of Composites Design, 2nd. Edition, Think Composites, Palo Alto, CA.

Manoochehr Zoghi,[1] Joel Sprague,[2] and Sam Allen[3]

Chapter 3

Emerging Geomaterials for Ground Improvement

Abstract

The age-old problem of land scarcity and the need to rebuild aging infrastructure in urban areas, increased realization of seismic hazards, and regulations mandated for various environmental problems, have been the impetus for the evolving ground improvement techniques during the past 25 years. Innovative ground modification approaches are routinely used now to solve unique geotechnical engineering problems, and often are considered to be the most economical means to improve an undesirable site condition. Following a brief overview of the relevant existing ground improvement materials and technologies, this chapter delineates the emerging geomaterials for civil engineering applications.

Keywords: ground, improvement, reinforcement, treatment, remediation, geomaterials, geosynthetics, admixtures, grouting

Introduction to Existing Technologies

In the past, when dealing with difficult sites for construction purposes, the conventional practice was limited to either replacing the unsuitable soils, or bypassing them with costly deep foundations. With the advent of ground modification techniques, in-situ soil improvement has become a viable alternative, and in most instances proves to be the most economical option where poor or unstable soils are encountered, especially on sites underlain by suspect or uncontrolled fills (Schaefer et al. 1997, Welsh 1997). Therein lies the principal objective of this chapter, to present an overview of existing technologies for soil improvement and stabilization, and to portray the emerging geomaterials for civil engineering applications.

---

[1] Associate Professor and Director of Civil Engineering Graduate Program, University of Dayton, 300 College Park Ave., Dayton, Ohio 45469.

[2] Senior Consultant, Sprague & Sprague Consulting Engineers, PO Box 9192, Greenville, SC 29604.

[3] Vice President, Geosynthetics Division, TRI/Environmental, Inc., 9063 Bee Caves Road, Austin, TX 78733.

The fact that many of the most suitable sites have been used, and there is an increasing demand for land reclamation and utilization of less favorable sites, has been the underlying reason for the continually evolving soil improvement techniques over the years (Fang 1992, 1995). Indeed, soil improvement and stabilization play a major role within the intertwining geotechnical, transportation, environmental, and hydraulics framework (Moseley 1993, Zoghi 1995). These ground improvement techniques comprise the mechanical modification, hydraulic modification, physical and chemical modifications, and modifications by inclusions and confinement. Detailed discussion of these technologies and their applications are delineated elsewhere (Hausmann 1994, Mitchell et al. 1992). The intent of this chapter is, however, to present the emerging geomaterials for soil improvement purposes. Specifically, the following applications will be considered herein: filtration and separation, stabilization and reinforcement, drainage and barrier provision, and erosion control.

Geosynthetics

Geosynthetics are a rapidly emerging family of geomaterials used in a wide variety of civil engineering applications (Koerner 1998). They are made almost entirely from polymers; although rubber, fiberglass, and natural materials are sometimes used (ASTM 1996). The American Society for Testing and Materials' (ASTM - D4439) definition of geosynthetics is: "A planar product manufactured from polymeric material used with soil, rock, earth, or other geotechnical engineering related material as an integral part of a man-made project, structure, or system." There are numerous styles and configurations of geosynthetics in the market; the predominant ones will be described in the ensuing sections of this chapter.

Knowing the primary function or performance characteristic of the geosynthetics facilitates the use of a "design-by-function" methodology as prescribed by Koerner (1998). Accordingly, geosynthetics may be used to function as a separator, filter, planar drain, reinforcement, cushion/protection, and/or as a liquid and gas barrier. Additionally, performance characteristics have been identified for newer geosynthetic applications. The various types of geosynthetics available, along with their specific applications, are discussed in subsequent sections. The relevant guidance on identifying the functions and performance characteristics associated with various applications and end uses of geosynthetics is tabulated below (Table 1).

Geotextiles – Almost all geotextiles available in the United States are manufactured from either polyester or polypropylene. Both polymers are synthetic "manmade" materials derived from the manufacture of crude oil and natural gas that have demonstrated a high level of durability, or resistance to degradation, in the soil environment.

There are two general categories of geotextiles: wovens and nonwovens. Woven geotextiles can be placed into two general categories - Slit Film and Filament. Woven geotextiles are made from either monofilaments, multifilaments, or slit films. There are two steps in this process: first, manufacture of the filaments or slitting the film; and second, weaving them to form the geotextile. Slit film wovens are manufactured by

Table 1 – *Functions*[1] *and Other Performance Characteristics*[2] *(ASTM 1994)*

CONTAINMENT[2] (C) A geosynthetic provides containment when it encapsulates or surrounds materials such as sand, rocks, and fresh concrete. (Note: during the placement of fresh concrete in a geotextile flexible form, the geosynthetic functions temporarily as a filter to allow excess water to escape.)

FILTRATION[1] (F) A geosynthetic performs the filtration function when the equilibrium geotextile-to-soil system allows for adequate liquid flow with limited soil loss across the plane of the geotextile over a service lifetime compatible with the application under consideration.

FLUID BARRIER[1] (FB) A geosynthetic performs the fluid barrier function when it essentially eliminates the migration of fluids through it.

FLUID TRANSMISSION[1] (a.k.a. drainage) (FT) A geosynthetic performs the fluid transmission function when the equilibrium geotextile-to-soil system allows for adequate flow with limited soil loss within the plane of the geotextile over a service lifetime compatible with the application under consideration.

INSULATION[2] (I) A geosynthetic provides insulation when it reduces the passage of heat, electricity, or sound.

PROTECTION[1] (P) A geosynthetic, placed between two materials, performs the protection function when it alleviates or distributes stresses and strains transmitted to the material to be protected.

REINFORCEMENT[1] (R) A geosynthetic performs the reinforcement function when it provides often synergistic improvement of a total system's strength created by the introduction of a tensile force into a soil (good in compression but poor in tension) or other disjointed and separated material.

SCREENING[2] (Scr) A geosynthetic, placed across the path of a flowing fluid (ground water, surface water, wind) carrying particles in suspension, provides screening when it retains some or all soil fine particles while allowing the fluid to pass through. After some period of time, particles accumulate against the screen, which requires that the screen be able to withstand pressures generated by the accumulated particles and the increasing fluid pressure.

SEPARATION[1] (S) A geosynthetic placed between dissimilar materials so that the integrity and functioning of both materials can remain intact or be improved performs the separation function.

SURFACE STABILIZATION[2] (SS) A geosynthetic, placed on a soil surface, provides surface stabilization when it restricts movement and prevents dispersion of surface soil particles subjected to erosion actions (rain, wind), often while allowing or promoting vegetative growth.

VEGETATIVE REINFORCEMENT[2] (VR) A geosynthetic provides vegetative reinforcement when it extends the erosion control limits and performance of vegetation.

[1] Functions – Functions are used in the context of this guide as terms, which can be quantitatively described, by standard tests and/or design techniques.

[2] Other Performance Characteristics – Other performance characteristics are qualitative descriptions, which are not yet supported by standard tests or generally accepted design techniques.

cutting sheets of impermeable film into narrow strips and then weaving these strips together to form a geotextile. These fabrics are normally used for sediment control, i.e. silt fence, and road stabilization applications. Their lack of permeability and irregular spaced openings make them poor choices for subsurface drainage and erosion control applications. Filament wovens, on the other hand, are manufactured by weaving extruded filaments (either single filaments or groups of filaments called multifilaments) together to form a geotextile. This process produces a geotextile with regular spaced openings, and this type geotextile is often used in erosion control applications. In recent years, high strength wovens (wide width tensile strength > 500 pounds/inch) have been added to the woven geotextile category. These geotextiles are very specialized and are primarily used in reinforcement applications. The physical properties of a woven geotextile can be controlled by varying the properties and spacing of the warp and filling yarns.

Nonwoven geotextiles are manufactured from either staple fibers or continuous filaments. The formation of both continuous filaments and staple fibers begins when the polymer is melted in a heated extruder and exits in a viscous liquid state. The polymer is funneled down branching pipes into extrusion dies with small holes. As the polymer flows out of these holes, it forms a continuous strand. The continuous strand can either be used to immediately construct a geotextile or "chopped" into short "staple" fibers and baled for later use. Continuous filament nonwovens are made by randomly distributing the extruded semi-solid continuous strands onto a moving belt in layers to form a felt-like "web". The web then passes through a needle loom and/or other bonding machine interlocking the filaments and giving the geotextile its characteristic tensile strength and dimensional stability. Upon exiting the needle loom the geotextile passes between powerful magnets and then through a metal detector to remove and, if not removed, detect broken needles. Subsequently, the geotextile is wound onto cores and cut into standard lengths. These final rolls are then sent for inspection, wrapping and storage. The staple nonwovens are made from short "staple" fibers that are typically 75-100 mm in length. Most nonwoven geotextiles available in North America are produced from staple fibers. The staple fibers are unbaled and dispersed onto a moving belt to form a web, which is bonded, usually by a combination of needlepunching and heat bonding.

It should be noted that the characteristics of a geotextile depend on the polymer type and the geotextile structure, including the method of bonding. For example, a nonwoven, needlepunched, continuous filament, polyester geotextile will have a uniform web with a randomly sized, 3-dimensional pore structure for more "soil-like" performance; multidirectional strength and conformability to resist installation damage and assure "intimate" soil contact; high strength to weight ratio to enhance cost-effectiveness; resistance to environmental degradation creep to facilitate extended design life. Alternatively, a woven, slit-film, polypropylene geotextile will have high bi-directional strength and stiffness to "bridge" local soft spots in a subgrade; light weight to provide ease of installation; and a high level of resistance to environmental degradation.

Geogrids – Geogrids are gridlike polymeric materials usually made from punching holes in and stretching sheets of high-density polyethylene or polypropylene or by weaving and coating high tenacity polyester yarns. The resulting grid structure possesses large open spaces (called apertures) ranging from 0.5 to 4.0 in. (1 to 10 cm), which enhance interaction with the soil. Their inherent high tensile strength and stiffness, or modulus, make geogrids especially appropriate for soil reinforcement.

Accordingly, the ASTM has defined the geogrid as: "A geosynthetic used for reinforcement which is formed by a regular network of tensile elements with apertures of sufficient size to allow strike-through of surrounding soil, rock or other geotechnical materials."

Geonets – Geonets are utilized exclusively for in-plane drainage. Nearly all the geonets are made of polyethylene. The molten polymer is extruded through slits in counterrotating dies, forming a tight net of closely spaced ribs (Koerner 1998). The ASTM - D4439 defines a geonet as: "A geosynthetic consisting of integrally connected parallel sets of ribs overlying similar sets at various angles for planar drainage of liquids or gases."

Geomembranes – There are two general categories of geomembranes: calendered and extruded. Calendered geomembranes are formed by working and flattening a molten viscous formulation between counterrotating rollers. Polyvinyl chloride (PVC), chlorosulfonated polyethylene (CSPE), chlorinated polyethylene (CPE), and, more recently, polypropylene (PP) are the most common calendered geomembranes. Specialty ethylene interpolymer alloy (EIA) geomembranes are used for unique applications.

Extruded geomembranes are manufactured by melting polymer resin, or chips, and forcing the molten polymer through a die using a screw-extruder. The sheet is formed either by a flat horizontal die or through a vertically oriented circular die to form either a flat wide sheet advanced on a conveyor belt, or cylindrical tube of "blown film", filled with air which is collapsed and pulled by nip rollers mounted high above the die. Blown film geomembranes must be slit prior to wind-up. Common extruded geomembranes include high-density polyethylene (HDPE) and various lower density polyethylenes, which are often grouped as very flexible polyethylene (VFPE). Polypropylene (PP) geomembrane is a relatively new type of extruded (as well as calendered) membrane, which is quickly gaining acceptance. Under development are fully crosslinked elastomeric alloys (FCEA) such as a thermoplastic elastomeric alloy of polypropylene and ethylene-propylene diene monomer (PP/EPDM).

Polypropylene (PP) geomembranes are made of a highly flexible form of polypropylene which is specially formulated to have all the benefits of an olefin, including chemical inertness to a wide variety of chemicals, while still being much more flexible and frictional than HDPE and not being susceptible to stress cracking. Finally the very Flexible Polyethylene (VFPE) is a new designation for the class of polyethylene materials, including VLDPE and LLDPE, which are formulated to be flexible like PVC.

Each type of geomembrane has unique characteristics relevant to its base polymer that leads to potential advantages and disadvantages as described in Table 2 (modified from Koerner, 1998). It is important to realize that site-specific decisions, relating to the use of geomembranes, should address the above issues initially regardless of membrane type.

As described below, variations in the manufacturing of geomembranes include the addition of reinforcing and/or texturing, coextrusion, and the availability of multiple thicknesses and sheet sizes.

Reinforcing – Calendered geomembranes often are manufactured with a woven polyester fabric sandwiched between two plys of polymer sheets resulting in a three-ply laminated geomembrane, which is sealed together, under pressure, to form a unitized system. A geomembrane which incorporates a reinforcing scrim is designated with an "R" accordingly, i.e. CPE-R, CSPE-R, PVC-R, EIA-R. Multiple plies are possible. The scrim imparts a major increase in mechanical properties over the unreinforced type, particularly in the tensile strength, modulus of elasticity and tear resistance of the final geomembrane.

Texturing – A textured surface can be created on either or both sides of a geomembrane by coextrusion with nitrogen gas, by impingement of polymer particles, or by lamination of polymer particles using foam. Surface texturing provides an effective means to enhance the interface friction between the geomembrane and adjacent soils or other geosynthetics.

Coextrusion – Two or three extruders simultaneously introduce molten polymer into the same die. As the different materials exit the die and are cooled they commingle with one another such that local blending and molecular entanglement occur and no discrete separation layer exists.

Thicknesses – Single ply geomembranes such as CPE, HDPE, PVC, VLDPE, LLDPE, PP can be manufactured up to 5 mm (200 mils) thick in a single sheet. Scrim-reinforced geomembranes such as CSPE-R and EIA-R are manufactured from single plies in the range of 0.25 to 0.38 mm (10 to 15 mils) which, when laminated, result in geomembranes of incremental thicknesses beginning at 0.85 mm (36 mils). It should be noted that most geomembranes used today are at least 0.50 mm (20 mils) thick and current minimum regulations for municipal solid waste (Subtitle D) liners require a geomembrane with a thickness of 0.75 mm (30 mils) or greater.

Sheet Sizes – Calendered geomembranes are typically 1 m to 2 m wide as they exit the calender and are transported in rolls weighing up to 680 kg to a panel fabrication facility. Large panels, commonly 5 to 10 rolls wide, are then factory fabricated on a project-specific basis. Panels are then accordion folded onto wooden pallets, covered and shipped to the job site for deployment. Factory fabrication of the largest panels that can be practically handled maximizes seam quality and facilitates efficient installation by minimizing the number of field seams, which must be made. Extruded geomembranes, on the other hand, are fed directly from the extruder to the wind-up core, creating full-width rolls weighing up to 2,270 kg. Typical roll dimensions will vary depending on geomembrane thickness with weight being the limiting factor. The 60-mil HDPE may come in relatively narrow rolls measuring 4.5 m by 338 m or significantly wider and shorter rolls that are up to 10.5-m wide and as short as 76 m. Generally, extruded geomembranes cannot be folded since they will develop permanent creases; therefore, all seaming is done in the field.

Table 2 – *Some Potential Advantages and Disadvantages of the Basic Polymers of Geomembranes*

POTENTIAL ADVANTAGES	POTENTIAL DISADVANTAGES
Chlorinated Polyethylene (CPE)	
Good weathering	Moderate Cost
Easy seams - dielectric & solvent	Plasticized with PVC
Cold crack resistance is good	Seam Reliability
Chemical resistance is good	Delamination possible
Chlorosulfonated Polyethylene (CSPE)	
Excellent weathering	Moderate Cost
Cold crack resistance is good	Fair in high temperatures
Chemical resistance is good	Blocking (sticking) possible
Good seams - heat and adhesive	
Elasticized Interpolymer Alloy (EIA-R)	
Excellent weathering	Moderate Cost
Cold crack resistance is good	Blocking (sticking) possible
Chemical resistance is good	Good seams - thermal
Good in high temperatures	Low thermal expansion/contraction
High Density Polyethylene (HDPE)	
Chemical resistance is excellent	Low-friction surfaces unless textured
Good seams: thermal and extrusion	Stress crack sensitive
Large variation in thickness	Seam workmanship critical
Low cost	High thermal expansion/contraction
Polyvinyl Chloride (PVC)	
Low cost	Plasticized for flexibility
Tough without reinforcement	Poor weathering, backfill req'd.
Lightweight as single ply	Plasticizer leaches over time
Good seams: solvent, heat	Cold crack must be verified
Large variation in thickness	High-temp. performance must be verified
Large variation in blends	Blocking (sticking) possible
	Chemical resistance must be verified
Very and Linear Low Density Polyethylene (VLDPE & LLDPE)	
Good seams: thermal and extrusion	Seam workmanship critical
Large variation in thickness	Chemical resistance must be verified
Low cost	Moderate thermal expansion/contraction
No stress crack	
Polypropylene (PP)	
Good seams: thermal and extrusion	Seam workmanship critical
Large variation in thickness	Limited field experience
Low cost	Moderate thermal expansion/contraction
No stress crack	
Chemical resistance is excellent	

Geosynthetic Clay Liners (GCL's) – An excellent overview of GCL's has been given by Daniel and Koerner, 1993; Daniel and Boardman, 1993; and Eith et al. 1991, from which the following information has been extracted. Geosynthetic clay liners are factory manufactured hydraulic barriers consisting of bentonite clay, supported by geotextiles and/or geomembranes, which are held together by needling, stitching and/or chemical adhesives. They are commonly used to augment or replace compacted clay liners or geomembranes. GCL's can be used as an independent hydraulic barrier to water, leachate or other liquids or along with a geomembrane in place of a compacted clay liner (CCL) to provide a "composite" barrier to these liquids.

In addition to the savings of over 3,822 m^3 of airspace per acre when compared to a three foot-thick compacted clay liner, the benefits of GCLs over traditional CCLs, include:

- Minimal labor required to install over 1 acre/day;
- Self-sealing, self-seaming upon hydration;
- Seams are simple overlaps (though sometimes additional bentonite is introduced into the overlap);
- Overlap guidelines are printed on each roll;
- Can be cut and trimmed with a utility knife to accommodate complex configurations;
- Bentonite rehydrates to negate the effects of freeze-thaw or desiccation;
- Lower installed cost (when clay is not readily available).

GCLs are manufactured in large, continuous operations by either placing bentonite clay and an adhesive in numerous layers on a "carrier" geotextile or geomembrane with an optional cover geotextile or geomembrane, or by placing layers of dry bentonite clay on a carrier geotextile and then covering with another geotextile prior to needlepunching or stitchbonding the entire GCL "sandwich". All of the GCL products in North America use sodium bentonite clay (sodium montmorillonite) powder or granules. This material occurs naturally in Wyoming, North Dakota and Montana in the U.S. The clay is mined, dried, pulverized, and sieved prior to shipment to the GCL manufacturing facility. Some GCLs used outside of North America, however, incorporate calcium bentonite. Calcium bentonite has lower water absorption and swell characteristics than does sodium bentonite. Additionally, the calcium bentonite can react with soda ash causing some of the calcium ions to be replaced by sodium ions and thereby producing a more absorptive, higher swelling bentonite. Similar ion exchange techniques can be used to modify or "engineer" bentonite to be more resistant to chemicals that are reactive with the naturally occurring clay, but this is done on a project specific basis working directly with the GCL manufacturer. At the present time, there are four major product types available on the North American GCL market. Each product type has its own unique properties. Additional products which use activated calcium bentonite have been developed and marketed in Europe and in Japan.

Furthermore, GCLs are factory made in widths of 4.1 to 5.2 m (13 to 17 ft) and lengths of 30 to 61 m (100 to 200 ft). Rolls generally weigh up to 1600 kg (3,500 lbs) and must be handled by specially outfitted equipment such as a tapered rod attached to the front of a fork lift or front-end loader, for loading and unloading, and by an appropriate support pipe and spreader bar for field deployment.

Geopipe – Another significant product which has been "adopted" as a geosynthetic is plastic pipe. The specific polymer resins used in the manufacturing of plastic pipes are: polyvinyl chloride (PVC), high-density polyethylene (HDPE), polypropylene (PP), polybutylene (PB), acrylonitrile butadiene styrene (ABS), and cellulose acetate buytrate (CAB). There is a wide variety of civil engineering applications for these products, including: highway and railway edge drains, interceptor drains, primary and secondary leachate removal systems. (Koerner et al. 1990).

Geofoam – A recent category of geosynthetic product is geofoam, which is the generic name for any foam material utilized for geotechnical applications. Horvath, being credited with introducing the name geofoam in 1992, has proposed the following definition (Horvath, 1995): "Geofoam is any manufactured material created by some expansion process that results in a material with a texture of numerous, closed, gas-filled cells." Although most geofoam materials are made of polymers, glass foam (cellular glass) has been and is still used. The most common type of polymer used in manufacturing of geofoam materials is polystyrene. Manufacturing involves either a two-stage molded-bead process of EPS (expanded polystyrene), or a continuous extrusion process of XPS (extruded polystyrene). The EPS are manufactured either as block molding with typical finished dimensions of 600 mm thick, 1200 mm wide, and 2400 mm long, or as custom shape molding producing specific shapes. The XPS are manufactured in plank or panel shaped pieces. Typical applications of geofoams include: as thermal insulations of road and airfield pavements and railway track systems, beneath on-grade storage tanks containing cold liquids, as lightweight fill, etc.

Geocomposites – The possibility of combining the superior features of various geosynthetics has rendered a host of so called "geocomposite" materials. Although there is an infinite number of possibilities in assembling different materials, only limited by one's ingenuity and imagination, the predominant geocomposite configurations consist of Geotextile-Geonet Composites, Geotextile-Geomembrane Composites, Geomembrane-Geogrid Composites, Geotextile-Geogrid Composites, Geotextile-Polymer Core Composites, and Geosynthetic-Soil Composites. It is beyond the scope of this manuscript to discuss the details of each of these products. An excellent overview of geocomposites is presented by Koerner (1998). The emerging geocomposite materials will be delineated in the ensuing sections.

Other Geomaterials

Other geomaterials used in civil engineering practice involve soil-admixtures, grouting, fiber-reinforced soils, soil nailing, and soil bioengineering materials. A brief review of the material characteristics and current applications will follow. The emerging issues will be discussed in the later sections of the manuscript.

Soil Admixtures – There are numerous admixtures used for various soil stabilization purposes; some of the more widely used materials are summarized below.

Soil-Cement – There are several types of cement and cementitious materials used for soil stabilization, the most common one being the ordinary portland cement. The process involves mixing the pulverized soil with cement in the presence of water and compacting the mixture to a high density. The treated material, known as soil-cement, possesses increased strength and stiffness, better volume stability, and enhanced durability. Three categories of cement-stabilized soils are prevalent: the compacted soil-cement, cement-modified soil, and plastic soil-cement. In the case of compacted soil-cement, sufficient cement is added to produce a hard and durable construction material, whereas, the cement-modified soil uses relatively small quantities of portland cement yielding an unhardened or semihardened mixture of soil and cement (Winterkorn et al. 1991). Although the end product of the plastic soil-cement is a hardened material, during placement, enough water is added to create a plastering mortar like material, which facilitates its application.

Soil-Lime – Two different forms of soil-lime stabilizations are commonplace in civil engineering applications: quicklime, and hydrated (or slaked) lime. The former is available in a coarse-grained powder with a bulk density of 0.85 to 1.05 t/m^3, while the latter is produced in the form of a fine powder with a bulk density of 0.45 to 0.6 t/m^3 or in a slurry form with a water content of 80 to 100%. The principal benefits include improved workability, increased strength, and volume stability (Broms 1993).

Soil-Bitumen – The use of bitumen, tar, and asphalt in ground modification is collectively referred to as bituminous stabilization. Bitumen is the byproduct of crude petroleum, whereas, the tar is the byproduct of coal or other carbonaceous material. The various bituminous stabilized products consist of soil-bitumen (or soil-asphalt)—a waterproofed cohesive soil possessing 4 to 7% bitumen, sand-bitumen (or sand-asphalt)—sand particles cemented together by adding 3 to 4% (sometimes as much as 10%) of a bituminous admixtures, waterproofed granular soil, a well-graded soil with some low plastic fines (waterproofed with 1 to 2% bitumen), and oiled earth - by spraying a bituminous emulsion or cutback on the soil surface which provides water and abrasion resistance. The bituminous admixtures, generally used to waterproof soils or occasionally reduce water absorption and add cohesion to granular soils, are most suitable for sandy gravels, sands, clayey and silty sands, and fine crushed rock (Winterkorn et al. 1991).

Soil-Power Plant Wastes – Pozzolanic mineral coal ash; i.e., fly ash, bottom ash, or FGD (flue gas desulfurization) sludge are frequently employed for soil treatment. Fly ash, a by-product of solid waste produced by the combustion of coal, is recovered from flue gases. It possesses a light to dark gray powder of predominantly silt-size particles.

Bottom ash, on the other hand, is coarser than fly ash (ranging in size from fine sand to gravel), and is obtained from the base of blast furnace. Combination of lime (L), cement (C), and flay ash (F) is also used in conjunction with aggregate (A) for improving the quality of pavement bases and subbases (Head et al. 1982).

Grouting – Grouting is the fluidized material, either in suspension or solution, injected into the voids of the ground or spaces between the ground and adjacent structures under pressure in order to produce stronger, denser, and less permeable soil, rock, or landfill (Coduto 1999, Welsh 1997a). Following are different grouting techniques.

Compaction Grouting – The ASCE's Geo-Institute Committee on Grouting has defined the compaction grouting as: "Grout injected with less than 25 mm slump. Normally a soil-cement with sufficient silt sizes to provide plasticity together with sufficient sand sizes to develop internal friction. The grout generally does not enter soil pores, but remains in a homogenous mass that gives controlled displacement to compact loose soils, gives controlled displacement for lifting of structures, or both" (ASCE 1980). Typical applications include limiting structural settlements, and for site improvement of deep, loose or soft soil strata.

Jet Grouting – Jet grouting is a technique where a cavity is excavated with a horizontal high-speed air/water jet emanating from a drill bit and then the cavity is filled with the appropriate grout. In this grouting process, the in-situ soil material is replaced with a stronger and/or less permeable grout-soil mixture. It can be used for creating cutoff walls, or cutoff barriers, underpinning, etc. (Kauschinger et al. 1992).

Chemical Grouting – This type of grouting utilizes properly formulated chemicals (based on sodium silicate solutions and inorganic reagents) to be injected into sandy soils (possessing less than 20% No. 200 fraction). It is often adopted for structural control in which the chemical grouting creates cohesion in the sand, making it similar to a sandstone. It is also used for water control in which a water barrier is created in granular soil formations (Bandimere 1997).

Slurry Grouting – The slurry grouting material generally consists of cement, bentonite, sand, additives, microfine cement, fly-ash, lime, and water which is injected into voids and cracks underground. Typically, it is used to reduce permeability of rock beneath new dams (Welsh 1987).

Fiber-Reinforced Soils – Construction of fiber-reinforced soils (FRS) deviate from the traditional soil reinforcing technique, in that the primary emphasis is on discrete, randomly oriented fibers as oppose to the continuous inclusions at a predetermined orientation. The shear strength of reinforced soil depends upon the fiber content—generally 3 to 4% by weight is the optimum quantity (Schaefer et al. 1997). Nevertheless the fiber inclusion increases the shear strength of both cohesive and cohesionless soils.

Soil Nailing – The basic concept of soil nailing is to reinforce the ground by closely spaced passive inclusions to enhance the overall shear strength (via frictional resistance at the interfaces) of the in-situ soil, thus, restraining its lateral movement (Chassie 1994). Generally, steel bars or other metallic elements are utilized along with the facing to construct a soil-nailing retaining system. In recent years, significant advances have been made for improving this technique, which will be discussed in the subsequent sections. It is worth noting that majority of the soil nailing projects to date involve temporary retaining structures. The substantial cost savings and reduced construction time are the attributes of soil nailing systems over the conventional rigid gravity retaining walls or external bracing systems.

Soil Bio-Engineering Materials – Soil bio-engineering, often considered as a subset of biotechnical stabilization, refers to the use of live plants and plant parts arranged in various configurations to serve as soil reinforcements, hydraulic drains, barriers to earth movement, and hydraulic pumps or wicks (Gray and Sotir 1995). A wide variety of biotechnical and soil bioengineering materials and methods are employed in construction practice. It is beyond the scope of this chapter to discuss the characteristics of the various materials/methods. An excellent treatise of biotechnical and soil bioengineering slope stabilization is given by (Gray and Sotir 1996). The emerging materials concerning the biotechnical and soil bioengineering stabilization will be presented along with other emerging geomaterials subsequently.

Prefabricated Vertical (Wick) Drains, PVD – Prefabricated Vertical Drains (PVD's) are often utilized to expedite the consolidation process, hence, to significantly decrease the settlement time of embankments over soft soils by providing a shorter drainage path. Other applications of the PVD's include: to reduce the pore water pressures due to seepage; to lower the perched ground water table; and to lessen the liquefaction potential in fine-grained soils (Schaefer et al. 1997). In general, PVD's are made of plastic cores, which are wrapped in geotextile filters to create channels to enhance the drainage (Koerner 1998).

Erosion and Sediment Control Materials – Geosynthetics, as well as natural materials, are used extensively in erosion and sediment control systems such as:

- Temporary, degradable materials for the enhancement of vegetative establishment;
- Long-term, nondegradable materials to extend the erosion control limits of vegetation, soil, or rock;
- Primary slope or channel linings;
- Silt fences and turbidity curtains;
- Components in sediment traps.

Erosion Control Materials – A large construction site may have several different erosion and sediment control materials depending on location (i.e., slopes vs.

channels), flow conditions (i.e., sheet vs. concentrated), and regulatory requirements.

Erosion Control Systems – There are two categories of erosion control systems: temporary, degradable and long-term, nondegradable. Temporary degradable systems include conventional loose mulches, as well as, hydraulic mulch geofibers (HMG), erosion control netting (ECN), open weave meshes (ECM), erosion control blankets (ECB), and fiber roving systems (FRS) (Table 3). The long-term systems include conventional sod and riprap, as well as, turf reinforcement mats (TRM), fabric formed revetments (FFR), geocellular confinement systems (GCS), gabions (G), and articulating concrete blocks (ACB). A brief description of each nonconventional erosion control system follows (Table 4).

Hydraulic Mulch Geofibers (HGM) – Longer lived, hydromulch is commonly used as an alternative to straw mulch. It consists of wood or paper fibers measuring less than 1/2" in length mixed in a tank of water (usually with seed). The mixture is then sprayed over the bare soil. As it dries it forms a thin mulch layer, yet it is still susceptible to the wash and wind blown problems associated with loose fiber mulches. West (1994) has reported on a very stable "bonded fiber matrix" (BFM) which overcomes many hydromulch drawbacks by incorporating a high strength adhesive which cross-links and is subsequently unaffected by wetting. Trotti (1994) describes the difference between BMF and normal hydroseeding techniques as "building a blanket" rather than just spraying mulch and seed.

Erosion Control Netting (ECN) – Lancaster and Austin (1994) describe erosion control netting (ECN) as 2-dimensional, woven natural fibers or geosynthetic, polyolefin biaxially-oriented process (BOP) nettings used for anchoring loose fiber mulches such as hay or straw. ECNs are rolled out over the seeded and mulched area and stapled or staked in place.

Open Weave Meshes (ECM) – Open woven meshes (ECMs) constructed of organic "twines" of jute or coir or polyolefin yarns provide an alternative to loose fiber mulches for erosion control. These meshes have gained wide use because of their continuity, uniform cover and hand installation. Organic meshes are 0.25 to 0.50 in. thick and typically have 1 in. or larger square uniform openings. Polyolefin meshes are considerably thinner with smaller openings. All meshes are very flexible, promoting intimate ground cover and, because of their woven structure, they are much more stable than loose fiber mulch. Organic meshes also absorb water, which can help maintain soil moisture. Still, the regular openings and 2-D structure do not provide for full coverage and the associated protection of the soil from rainfall impact and full area seed coverage. Polyolefin and jute meshes degrade within 12 months, while coir may last 2 to 3 times as long.

Table 3 – *Some Potential Advantages and Disadvantages of Temporary, Degradable Erosion Control Systems*

POTENTIAL ADVANTAGES	POTENTIAL DISADVANTAGES
Loose Mulches	
Lowest cost Well accepted High installation rate Moderate sediment yield Moderate vegetative density	Very temporary (< few weeks/months) No concentrated flows Dusty Requires anchoring (i.e. crimping, tackifier) May require certification (i.e. noxious weeds)
Hydraulic Mulch Geofibers (HMG)	
Low cost Well accepted High installation rate Moderate sediment yield Moderate vegetative density	Very temporary (< few weeks/months) Very low or no concentrated flows
Erosion Control Netting (ECN)	
More effective than tackifier (if properly anchored)	Temporary (1-2 yrs.) More costly than tackifier Netting may interfere w/ maintenance (if inadequately anchored)
Open Weave Meshes (ECM)	
Low to moderate cost Moderate sediment yields Moderate vegetative density Moderate moisture absorption	Temporary (1-2 yrs.) Low flows only Not complete ground cover
Erosion Control Blankets (ECB)	
Low to moderate cost Easy to install Good moisture absorption Very low sediment yield Very good vegetative density	Temporary (1-3 yrs.) Low to moderate flows Netting may interfere with maintenance (if improperly manufactured or inadequately anchored)
Fiber Roving Systems (FRS)	
Low to moderate cost High subgrade conformance Very low sediment yield Very good vegetative density	Temporary (1-2 yrs.) Low to moderate flows only Special equipment required

Table 4 – Some Potential Advantages and Disadvantages of Long-term, Nondegradable Erosion Control Systems

	POTENTIAL ADVANTAGES	POTENTIAL DISADVANTAGES
Sod		
	Immediate vegetation and its associated benefits	High costs Risk of dying before establishment May need irrigation Limited to turf grasses (not native species)
Turf Reinforcement Mats (TRM)		
	Moderate costs Long-term (indefinite) Moderate to high flows Encourages infiltration Mod. To high vegetative density Extends the performance limits of vegetation Flexible – conform to differential settlement	Low to moderate sediment yields (unvegetated) Requires vegetative establishment for effective long-term performance
Riprap		
	Long-term (requires maintenance) Moderate to high flows	Moderate to high cost May have negative aesthetic & safety impact
	Encourages infiltration Low to moderate sediment yield (when underlain by a geotextile)	Difficult to install correctly on steep slopes Low vegetative density Often unstable, especially w/o geotextile
Other "Hard" Systems Including Fabric Formed Revetments (FFR); Geocellular Confinement Systems (GCS); Gabions (G); Articulating Concrete Blocks (ACB)		
	Long-term (indefinite) Moderate to Very high flows Low to moderate waves Low to moderate sediment yields Can accommodate various fill materials Durable and relatively maintenance free	High to very high costs No to delayed vegetation establishment No to low vegetation density May prevent infiltration Special deployment/equipment requirements

Erosion Control Blankets (ECB) – Carroll et al. (1992) considered erosion control blankets to be a technological improvement over both loose fiber and open weave meshes (ECMs). ECBs are organic fiber-filled blankets consisting of straw, wood (excelsior), or coconut fibers sewn to or between synthetic nettings. ECBs provide a thick (>0.5 in.) full coverage of mulch which better absorbs rainfall impact and retains moisture. The synthetic nettings add strength to help ECBs resist tractive forces associated with overland flow. Still, their useful life is limited to 12 to 24 months.

Fiber Roving Systems (FRS) – Agnew and Thiesen (1994) describe the construction of fiber roving systems using fibrillated, or split, fibers fed off spools and applied through a specially designed air gun. Using compressed air, the roving is continuously fed from the spools through the air gun and uniformly applied over a seeded soil surface. A randomly cross-linked mat of continuous strands results and provides a high percentage of ground cover. A tackifier is then sprayed on top of the FRS to firmly anchor it to the soil. FRS can conform to almost any geometry.

Turf Reinforcement Mats (TRM) – Turf reinforcement mats (TRMs) are 3-dimensional structures composed of fused polymer nettings, randomly laid monofilaments, or more recently yarns woven or tufted into an open, dimensionally stable mat. The dimensional stability produces a somewhat stiffer, but much stronger mat. According to Theisen and Carroll (1990), these flexible, synthetic, 3-dimensional mats are designed to be used in conjunction with topsoil and seed or turf to create strong, durable and continuous soil-root-mat matrices which can provide nearly twice the erosion protection of plain grass alone.

Fabric Formed Revetments (FFR) – Sprague and Koutsourais (1992) describe the construction of fabric formed revetments (FFR) as taking advantage of low-cost, durable, synthetic fabrics to produce 3-dimensional forms for casting concrete slabs. By pumping a very fluid fine-aggregate grout into a fabric envelope consisting of 2 layers connected by tie-chords or by interweaving at points, a concrete mattress can be constructed in minutes. FFRs provide the durability of rigid linings, such as cast-in-place concrete or asphaltic concrete, and the flexibility and/or water permeability of protective rock systems such as riprap or gabions.

Geocellular Confinement Systems (GCS) – Geocellular confinement systems (GCS) are 3-dimensional honeycomb-like structures filled with soil, rock or concrete. The GCS structure, often called a geocell, is made of strips of polymer sheet or geotextile connected at staggered points so that, when the strips are pulled apart, a large honey-comb mat is formed. Geocell thickness (depth) ranges from 2 in. to 12 in. According to Wu and Austin (1992), a GCS is effective at assuring that surface soil is retained on a slope. The GCS provides both a physical containment of a depth of soil and a transfer of load through the GCS. The cell walls confine the soil, intercept surface water flow and, in so doing, decrease runoff and increase soil moisture.

Gabions (G) – Gabions are compartmented rectangular containers made of galvanized steel hexagonal wire mesh and filled with hand-sized stone according to Theisen (1992). Cells of equal capacity are formed by factory-inserted wire netting diaphragms or partitions, which add strength to the container and help maintain its shape during the placement of stone. In highly corrosive conditions a polyvinyl chloride coating is used over the galvanized wire. Gabions are more flexible, durable and permeable than rigid structures and more stable than loose rock. Additionally, as sediment collects in the stone fill, plant growth can be supported.

Articulating Concrete Blocks (ACB) – Articulating concrete block (ACB) revetments are described by Koutsourais and Sprague (1992) as interlocking cellular concrete blocks, with varying amounts of open area within or between blocks, underlain by a filtration geotextile. The blocks are held on the slope by anchors placed at the top of the slope and/or friction between the slope and the blocks. The blocks can be assembled into mats either at the factory or on site. Sections or mats of precabled ACBs can be placed by using a special spreader bar. ACBs may also be placed by hand on-site with or without cables. ACB revetments provide the porosity, flexibility, vegetation encouragement and habitat enhancement, and ease of installation of rolled GECs and the nonerodibility; self-weight, and high shear resistance of rigid linings. Some of the potential advantages and disadvantages are tabulated herein.

Those erosion control systems, which are premanufactured, rolled up, and shipped to the site for installation, are often called rolled erosion control products, or RECPs. Typically, temporary degradable RECPs are manufactured in 1.9 m wide and up to 26 m long rolls with handling being the practical limitations. On the other hand, long-term, nondegradable RECPs come in relatively narrow rolls measuring 0.9 m by 152.4 m or significantly wider and shorter rolls that are up to 2.3 m wide and as short as 27.5 m. Non-RECPs have unique modular dimensions and are commonly joined in the field to create expansive mats.

Sediment Control Systems – In general, the benefits of geosynthetic sediment control systems over traditional structures include: minimal labor required to install; low cost; highly efficient in removing sediment; very durable and sometimes reusable. Theisen (1993) has summarized the performance of a well-designed silt fence as follows:

- It initially screens silt and sand particles from runoff.
- A soil filter is formed adjacent to the silt fence and reduces the ability of water to flow through the fence.
- This leads to the creation of a pond behind the fence, which serves as a sedimentation basin to collect suspended soils from runoff water.

To perform satisfactorily as a silt fence, the geotextile must have properly sized openings, which initiate the formation of a soil filter. Also the strength and storage capacity of the fence must be adequate to contain the volume of water and sediment anticipated during a major storm. Hydraulic loading during a storm event is commonly the primary loading considered.

Other sediment control systems include: turbidity curtains and porous sediment control structures. Turbidity curtains are reusable floating panels that prevent water-polluting sediment from shore-side construction or off-shore filling and dredging operations from moving off-site. The top edge of each curtain contains floats and a cable or chain. Weights are attached to the lower edge of the curtain to keep it vertical in the water. Posts, piling, or anchors hold the curtains in place. Generally, they are intended for use with currents no greater than about 1.5 m/s and to depths of no more than 4.6 m to 7.6 m.

Theisen (1993) reported on the development of porous sediment control structures. These are three-dimensional moldable masses of crimped polypropylene or other type fibers, which are placed in rills or gullies to provide passive sediment control. They are arranged by hand with size and shape determined by the installer.

Governing Regulations & Standards of Practice

Much of the development of geosynthetics technology has been in response to government regulations and/or as the result of much government study. Environmental applications have been especially impacted by regulations while the transportation sector has found —substantially through government initiatives—that geosynthetics are much more cost-effective than many conventional technologies.

Environmental Subtitle C & D – From the mid 1980s to the early 1990s, the Environmental Protection Agency (USEPA) recognized the benefits of geosynthetics in the protection of groundwater and issued regulations (Subtitle C & D) requiring, or at least accepting, geosynthetics in landfill lining and capping systems. Subsequently, geomembranes and GCLs have become standard liner components while geotextiles and geonets are commonly used in leachate, gas, and infiltration water collection systems. Additionally, it is not uncommon for geogrids and geotextiles to be used for reinforcing subgrades or soil veneers in landfill construction.

Transportation / AASHTO M288 & Bridge Specs – According to Sprague (1998), since 1986, the use of reinforced soil systems, which are also called mechanically stabilized earth (MSE), in the United States has significantly increased. The primary types of MSE systems, which have emerged, include: mechanically stabilized earth walls (MSEW); reinforced soil slopes (RSS); reinforced embankments over soft foundations (RESF); and reinforced liners and caps. MSEW and RSS have become especially important in highway construction as their use reduces the required width of new right-of-way and facilitates construction within existing limited right-of-way. RESF are recognized as a cost-effective alternative to traditional techniques for constructing earthen embankments over low strength foundations. Additionally, critical environmental applications of geosynthetic reinforcement have emerged, including the reinforcement of landfill and lagoon liner and cap systems on steep slopes and over weak subgrades or underlying voids.

Clearly, there has been substantial acceptance of geosynthetic reinforced soil structures and much of the growth has been the result of aggressive marketing by companies providing candidate geosynthetic soil reinforcement materials. Therefore, much of the experience gained to-date is with those reinforcement materials, which are marketed most aggressively and "packaged" with proprietary designs. Yet, this growth has been fraught with argument as competing proprietary systems claim superior performance or dependability. In response, independent researchers and Federal and State engineers have advanced the development of generic design techniques and material specifications such as AASHTO (1997) MSEW specifications and the FHWA's RSS guidelines by Berg (1993) and the DEMO 82 educational effort discussed by

DiMaggio (1996). This new generation of generic design techniques and specifications recognizes the suitability of both geotextiles and geogrids for soil reinforcement.

The FHWA has taken the lead in providing guidance in the implementation of generic design techniques and material specifications associated with geosynthetic reinforced soil systems. Still, individual owners/agencies are free to exercise their own judgements as to the appropriateness of following the FHWA lead. Sprague (1998) presented the results of a survey of state Departments of Transportation (DOTs) conducted in the fall of 1996. The survey solicited responses on the use of geosynthetic reinforced soil structures, reinforcement preferences, and reasons why preferential specifications may be used on any given project. The following trends were apparent:

1. Most Departments of Transportation (DOTs) are commonly using at least one form of geosynthetic reinforced soil structure.
2. Geogrids have been the most often used geosynthetic reinforcement to-date, largely because of vendor/consultant influences and greater experience with these materials.
3. Many DOTs intend to use non-preferential, more generic design/specification approaches in the future.

NPDES/SWPPP– A progression of regulatory actions have brought a national focus on erosion control, including:

- Amendments to the Federal Water Pollution Control Act (1985) - elimination discharge of any pollutant to navigable waters.
- The Clean Water Act (1987) - requiring National Pollution Discharge Elimination System (NPDES) Permits for construction activities.
- Intermodal Surface Transportation Efficiency Act (1991) - requiring erosion control guidelines for all federal-aid construction projects. This lead to AASHTO's "Erosion and Sediment Control in Highway Construction," in Volume III, Highway Drainage Guidelines (1992). This has subsequently been made a regulatory document.
- Coastal Zone Act Reauthorization Amendments (1990) - requires measures to control non-point sources of pollution in coastal areas.

The centerpiece of these regulatory actions is the NPDES permitting process, which is required for construction activities. Coverage under a state's general NPDES permit requires:

- Information about the construction activity
- Storm Water Pollution Prevention Plan
- A Notice of Intent (NOI) to be covered

The Storm Water Pollution Prevention Plan (SWPPP) must include:

1. The technical basis used to select the pollution control practices to avoid increasing the historical amount of sediment in water.
2. The maintenance of each sediment and erosion control measure.

Emerging Issues

The ground improvement techniques have continued to grow over the last two decades. Many techniques, developed over the years are now used routinely such as various types of grouting. Still new materials and methods are being developed, tried, and applied to different situations such as vitrification and sparging. In addition to the traditional ground modification applications, they are now being used to mitigate potential problems associated with liquefaction (Mitchell, Baxter, and Munson 1995), water flow, and environmental fields for cleanup and isolation of contaminated sites. Indeed, there are many innovative techniques being invented, researched, and experimented in conjunction with environmental remediation industry, bound only by available funds and the willingness to try (Schaefer et al. 1997).

As far as the soil reinforcement is concerned, its use over the last two decades has matured this technology. Nevertheless, the emergence of any new material or technology related to soil reinforcement in the future will be incremental and modest.

Durability – How long will they last? This question has been asked of geosynthetics for more than three decades. Thankfully, both experience and extensive research have demonstrated that, in the typical soil environment, geosynthetics have a performance life far exceeding 100 years. The FHWA has extensively researched geosynthetic durability and developed prequalification criteria for durable geosynthetic polymers.

Economics – As technologies mature, one clear benefit is the cost-effectiveness brought on by the greater competition that follows increase demand. This benefit is certainly becoming a reality for geosynthetics. Their use is now "standard practice" in many applications and, as a result, their price has trended downward.

The widespread use of geosynthetics and the associated expanding supply has involved more professionals in the development of geosynthetics technology. This interest holds the promise of making technological advancements less costly and quicker to market.

Proprietary Restrictions – Contrary to the trends noted in Section 1.4.2, proprietary interests can impact the emergence of geosynthetics. As with any product, patents are often sought for geosynthetic innovations in an attempt to afford the inventor greater opportunity to benefit from his invention. These proprietary protections can cause some evolving geosynthetic technologies to be expensive and slow their acceptance. But, the benefits of most of these geosynthetic innovations have proven to out-weigh the extra cost associated with their proprietary protections.

Emerging Filtration & Separation Technologies

Geosynthetics

Geosynthetics' use in filtration and separation technologies has become standard of practice and standard specifications have been developed by most transportation agencies to facilitate their use. As a result, there has been little demand for innovation in these areas. Still, some modest, niche development efforts have been undertaken as discussed below.

Geotextiles – Some landfill designers have been concerned about the build-up of microorganisms adjacent to the filtration geotextile causing clogging. In response, geotextiles incorporating a biocide have been developed but are not widely used.

In the area of separation of coarse and fine materials, as in road base construction over water-sensitive subgrades, the trend has been toward using slightly more massive (increased mass per unit area) geotextiles to provide greater resistance to puncture during construction.

Other Geomaterials

Proprietary Drains – Hansbo (1993) presents the overall features of different proprietary drains. He explains that all types of filters are manufactured of synthetic material, some of which are made in the shape of a loose sleeve enclosing the core while others are fixed to the core. Nevertheless, one needs to evaluate how the drain core and filter affect the efficiency and reliability of that drain when considering the emerging products.

Emerging Stabilization & Reinforcement Technologies

Geosynthetics

Stabilization generally refers to enhancing support of dynamic loads related to construction of roads over soft soils. Recent studies and public discussions have attempted to address the relative benefits of geotextiles (providing separation and membrane reinforcement) and geogrids (providing confinement and membrane reinforcement) in road stabilization. Reinforcement relates to increasing the geotechnical stability under generally static loading. Geosynthetics have a long and successful, but relatively infrequent history in soil reinforcement. Acceptance of geosynthetics in stabilization and reinforcement has grown dramatically and several trends in technology are now clearly identifiable.

Free draining granular material is the preferred backfill material in reinforced soil structures. Therefore, the use of reinforced soil structures - primarily MSEW and RSS - has been somewhat limited in cases where high quality backfill material is not readily available. Mitchell and Zornberg (1994 & 1995) observed that the inherent low strength, moisture instability, possible volume changes, and creep potential, along with the corrosive potential of saturated soils, are common concerns associated with the use of fine-grained, cohesive backfills. They further noted that as successful experiences are documented and consistent design methodologies are developed for these poorly draining backfills, even greater growth in the use of reinforced soil structures will be possible.

Geotextiles – Road stabilization using geotextiles has largely been focused on unsurfaced roads, since significant straining of the geotextile is necessary to mobilize geotextile strength and this straining will only happen when rutting occurs. Stabilization of unsurfaced roads with geotextiles was among the earliest uses of geotextiles, and the application has been well studied. Therefore, most innovation has centered on the definition of high-strength geotextile durability properties to assure lifetimes of 75 to 100 years as well as the combining of a thick nonwoven geotextile with a high-strength geotextile to provide both reinforcement and drainage (for consolidation) in fine-grained soils.

Geogrids – Geogrids provide reinforcement along with enhanced soil interlock. Their stiffness and open apertures makes them especially compatible with segmental retaining wall construction. Consequently, the demand for geogrids has grown with the acceptance of segmental retaining walls for a wider range of applications, including highway retaining walls and bridge abutments. Though HDPE and PP geogrids have long dominated the market place, in recent years, several coated polyester (PET) geogrids have been introduced as companies attempt to capitalize on PET's strength, stiffness, and creep resistance.

Geogrid stiffness and ability to interlock with road base aggregate to create a "confining" phenomena within the base material is often argued as a unique (but as yet unmeasured) feature of geogrids in road stabilization. Development of testing techniques to measure confinement properties is underway. Identification of confinement properties is critical to developing generic design techniques for including geogrids in paved road designs.

Soil Fibers – The preceding techniques of soil reinforcement primarily use continuous inclusions, in predetermined orientation, to achieve the optimum strength and stability in a soil mass. Accordingly, the soil reinforcement elements including, all fabrics, grid, and strip are laid out sequentially in alternating layers (Schaefer et al. 1997). Emerging techniques of soil reinforcements comprise discrete, randomly oriented fibers. Leflaive (1985) is credited with using single monofilament fibers, injected randomly into sand (designated as "Texol"), for reinforcement purposes. McGown et al. (1985) developed randomly distributed polymeric mesh elements, oriented and randomly distributed discrete fibers for effective reinforcement of both cohesionless and cohesive soils.

The behavior of oriented fiber reinforced soils (FRS) under static loads has been investigated by numerous researchers (Gray and Ohashi 1983, Jewell 1980). It has been found that the oriented fibers; (a) increase the peak shear strength, while limiting the post peak drop off; (b) have no effect on the frictional properties of sand, in essence, the

Mohr-Coulomb failure envelope remain parallel with the unreinforced envelope above a minimum confining stress; (c) at inclination of 60° will coincide with the direction of maximum tensile strain, providing an optimum orientation for maximum fiber contribution to shear strength; d) with 90° initial inclination is a good approximation for configuring the contribution of randomly distributed fibers.

In contrast to the continuous, oriented FRS as described above, there is a limited data available with regard to the performance of discrete, randomly distributed fibers (DRDF). Gray and Al-Refeai (1986) contend that the engineering properties of a soil mass reinforced with DRDF are similar to the traditional reinforced soils, however, DRDF reinforced soils resemble that of admixture stabilization (similar to other soil stabilizers such as cement, lime, etc) for their preparation. Accordingly, the soil is mixed with DRDF and then compacted in the field utilizing either standard or modified compaction methods. The inherent advantage of DRDF over FRS is the maintenance of strength isotropy and the absence of potential planes of weakness that exist parallel to oriented reinforcement (Maher and Gray, 1990, Maher and Woods 1990). Other notable emerging materials for soil reinforcement include discretely-shaped recycled and waste products such as shredded tires and fibrous by-products of manufactured glass and plastics (Foose et al. 1996).

The emergence of engineered fibers and recycled fibrous materials for soil reinforcement applications, is partly due to shorter construction time and lower cost (Schaefer et al. 1997). The impact of FRS and DRDF can be summarized as: (1) increase in the shear strength of both cohesionless and cohesive soils significantly; (2) the latter technique renders strength isotropy in a soil mass with absence of potential planes of weakness; (3) depending upon the fiber content (3 to 4% by weight is considered to be as optimum content), fiber length or aspect ratio (longer fibers in sand and shorter ones in clay), fiber stiffness and surface friction, the strength is increased; (4) furthermore, higher strength can be anticipated for well graded and angular sands. Another emerging application is related to dynamic loading; specifically, during seismic events. The inclusion of fibers in susceptible soils, has been found to enhance resistance to liquefaction and increase shear modulus of sand both under low and high amplitude vibrations.

Other Geomaterials

As discussed earlier, the soil improvement technology has come of age during the past decade, and in many respects it is considered fully developed. Soil improvement is now considered as a viable alternative to the conventional methods routinely and enjoys a mature marketplace. Ground treatments are generally utilized to strengthen the soil, decrease the compressibility/swell potential, and reduce the permeability, which in turn yields a more cost effective construction. Additionally, new methods are being developed, researched, and applied to the environmental field to assist in the cleanup and isolation of contaminated sites (Schaefer et al. 1997).

Soil Admixtures – Cement is generally considered to be the best type of admixtures, producing a relatively high strength stabilized soil as compared to other types of soils. One of the most recent development concerning the cement-stabilized soils is the work of Bergado et al. (1996) for improving the properties of weak Bangkok clay. In this research study, it was found that mixing ten to fifteen percent cement (by weight) and curing time of one to two months can be considered as the optimum in enhancing the unconfined compressive strength and preconsolidation pressure.

Recent developments of bentonite-soil admixtures have centered on landfill liners, covers, and cut-off walls. Anderson and Hee (1995) used bentonite admixture to bring the lateritic soils into the acceptable range of hydraulic conductivity in conjunction with landfill construction. The results of their study revealed that a significant difference exist between excavated soils and stockpiled soils. It was further determined that five percent bentonite is needed to achieve adequate compaction for an acceptable range of permeability at the confining pressure of 10 kPa.

Fly ash, a by-product from the combustion of ground and powdered coal used for power generation, is susceptible to moisture and density. There has been a continuous effort in the research and development of basic technology for treating soils with fly ash during the past two decades. The emerging technologies, however, entail the use of fly ash for the solidification and disposal of industrial wastes, sludge, and slag. Another potential application is the use of lime sludge amended fly ash as an engineering material in conjunction with landfill caps, liners, and engineered fill (Maher et al. 1993). This same study also revealed that the combination of fly ash and lime sludge compacted well and possessed relatively high compressive strength when exposed to air-drying over a short period of time. Another recent research study, conducted by Khan and Saker (1993), adopted a mixture of fly ash and enzymes to further improve the strength and stability of soils. These researchers determined that the addition of merely 5 percent enzyme to the mixtures will increase the unconfined compressive strength by two fold.

Grouting – State-of-the-art report along with recent technology advances, engineering applications and research results concerning various types of grouting are presented in an ASCE Geotechnical Special Publication (Vipulanandan 1997). The information presented here are mostly taken from that publication.

As indicated in Section 1.2.2, there are four basic types of grouting. According to Welsh (1992), the slurry, compaction, and chemical grouting are matured systems in the United States, whereas the jet grouting is considered as an emerging technology (Tarricone, 1994). Another type of grouting, which was developed overseas and was recently introduced into the United States is known as fracture grouting (Welsh 1992).

Bandimere (1997) points out that "As with any emerging technology, there are numerous spin-off applications, that while related and valid in their own right, may not necessarily be true to the original concept." A case in point is a patented process of injecting waste into waste and controlling the end product, which is considered as one of the emerging slurry grouting type of technologies in the 90's (Chesner and Welsh 1992).

Al-Alusi (1997) explains that compaction grouting, a method of in-situ soil densification by grout injection under pressure, emerged from "mudjacking" applications, through deep slurry consolidation to evolve into a compaction procedure, which was " perpetually perfected using available theory to reach today's state-of-the-art technology." Among the more recent innovative applications, hence, emerging concept in the field of compaction grouting is building continuous foundation element, i.e., the grout pile (Schaefer et al. 1997). The construction procedure of this emerging technology is based on pressure-injected footing at required depth to support a column of compaction grout to desired elevation. Although continuous reinforcement is often incorporated into the grout pile, however, according to the ground improvement committee, there is a need for future research on the reinforcing aspects of the compaction grout bulbs or columns formed in the ground with regard to the composite properties of the treated soil mass (Schaefer et al. 1997). It is postulated that this endeavor will permit even more applications in solving geotechnical problems.

Most recently, Pearlman (1998) cited numerous emerging trends concerning the jet grouting applications. These include: "use of jet grouting to create preinstalled structural supports for cut-and-cover excavations in soft ground and pervious soils. Also, preinstalled jet grouting supports are seeing greater use because of the need to conform to the growing number of environmental regulations, meet the risk of dewatering in urban areas and deal with unexpected round conditions."

Emerging Drainage Technologies

Geosynthetics

Geonets – In response to a common need for increased flow in drainage layers associated with both leachate collection in liner systems and infiltration collection in capping systems, multi-layer geonet systems have been introduced. These manufactured multi-layer systems provide increased flows without introducing the potential for a slip plane or compressive collapse within the drainage layer.

Geocomposites – Prefabricated drainage composites are evolving into pre-engineered systems offering unique combinations of plastic core and filtration geotextile to meet project-specific requirements. Additionally, an impermeable geomembrane and/or cushion material can be attached to the geocomposite to facilitate foundation waterproofing and drainage to be constructed with one product. An alternative, low cost approach to foundation drainage is to use a cushion layer over the foundation waterproofing which will double as a drainage layer. This "secondary" attention to drainage is only appropriate in fine-grained soils. Drainage collection and discharge is also very important and has been enhanced by the development of larger capacity drains and associated fittings, which are more easily installed than are traditional pipe and gravel systems.

Prefabricated Vertical Drains (PVD) – The emerging applications of PVD are considered to be in the stabilization of hazardous waste and to reduce potential for liquefaction. Successful implementation of PV drains to stabilize water tables and to help in providing a vapor extraction system have been cited recently (Schaefer et al. 1997) On one hand, the PV drains enhance the drainage, thus offer a pore pressure relief mechanism. On the other hand, some tensile strength is provided due to reinforcement characteristics of PVD, which will reduce the liquefaction potential. It is anticipated that in the next decade numerous applications of PV drains for both reduction of liquefaction potential and to stabilize the hazardous chemical sites will be accomplished.

Other Geomaterials

Natural Prefabricated Vertical Drains (PVD) – The benefits of PVD technology with a "natural" twist has evolved in some developing countries. Specifically, natural PVDs have been developed which use a coarse 3-dimensional weave of large denier coir or other natural yarns as the core and a filter envelope of finely woven small denier jute or other natural yarns.

Emerging Barrier Technologies

Geosynthetics

Geomembranes – Recent innovations in geomembranes are designed to address critical barrier concerns, including the identification of leaks and the elimination of excess expansion and contraction. These concerns are being addressed by the introduction of conductive and co-extruded geomembranes, respectively. Conductive geomembranes have electrically conductive materials embedded in them, which can detect the presence of excess liquid beneath the liner when a charge is induced. Both size and location of the leak can be detected. Co-extruded geomembranes incorporate a light colored surface to reflect sunlight and, therefore, to reduce heat generation and associated expansion in the geomembrane.

Another important, and on-going, innovation in geomembranes is the on-going development of new polymer formulations to in-service performance and/or ease of installation. The most significant recent membrane formulation is the introduction of flexible polypropylene geomembranes. PP geomembranes hold the promise of both high chemical resistance (similar to HDPE) and flexibility (similar to PVC).

"Structured" geomembranes are another interesting innovation. These geomembranes have "structures" manufactured on their surface, such as "cleat-type" projections to enhance interface friction with adjacent soil or "columns" to create a drainage space between the membrane and an overlying geotextile, eliminating the need for a separate geonet. More structured geomembranes can be expected as manufacturers address designers' needs.

GCLs – Geosynthetic clay liners (GCLs) are now well recognized for their ability to provide very low permeabilities and self-healing properties, but with the critical concern of low internal shear stability associated with bentonite. Recent innovations in manufacturing have incorporated both needlepunching and surface melting of the protruding needlepunched fibers to "reinforce" the thin bentonite layer allowing the reinforced GCLs to be used confidently on steeper slopes.

Vertical Barriers – Remediation schemes associated with contaminated sites requiring lateral containment of groundwater have benefited recently by the introduction of durable geomembranes as vertical barriers or cutoff walls. Thick HDPE geomembranes are fabricated with interlocks along the vertical edges which, when incorporating a water-expanding material, provide a more dependable, continuously watertight seal than do traditional cut-off techniques. Often these vertical barriers are installed in a bentonite slurry trench to enhance installation and long-term performance.

Other Geomaterials

Bentonite Mixing – The recent (and emerging) applications of the bentonite-soil-mixing encompasses soil landfill liners, covers, and cutoff walls in order to increase their flexibility and decrease the permeability. Further more, bentonite can be mixed with other additives, such as cement. It has also been used for dam cutoffs.

Spray-on Systems – One way to eliminate the huge number of geomembrane seams which must be checked and tested when installing a liner is to make the liner seamless. Spray-on technologies, which use a geotextile substrate, continue to hold promise in this area. Yet, substantial work is necessary to bring down the cost of spray-on systems and to enhance the ability to monitor barrier thickness. Spray-on systems can also be formulated with greater chemical resistance for unique, demanding environments.

Leak Detection – One never knows how successfully a liner has been installed. Extensive field quality control and quality assurance help substantially reduce the chances of leakage, but only leak detection can prove that a complete seal has been obtained. Highly scientific, yet amazingly straightforward techniques are being developed which can detect and pinpoint leaks. This kind of quality assurance is expected to continue to advance to the point where little, if any, destructive sampling of seams will be justified.

Emerging Erosion and Sediment Control Technologies

Geosynthetics

Rolled Erosion Control Products (RECPs) – RECPs are designed to nurture and support vegetation. Many different RECPs have emerged in recent years, which have enhanced erosion control and extended the ability of vegetation to resist long-term erosion forces. What is just now emerging is an understanding of why these products work so well. The industry is developing performance tests designed to quantify the relative effectiveness of various RECPs. These tests will help optimize the performance and cost of RECPs.

Silt Fences – Silt fences are primarily used to prevent topsoil from migrating off to construction site; to protect adjacent property from siltation; and to protect fish and other wildlife from the siltation of ponds, lakes, streams, and rivers (Shea 1996). An emerging application of silt fences is to mark or delineate a wetland area that is not to be disturbed (Shea 1996).

Other Geomaterials

Gray and Sotir (1996) have summarized the State-of-the-Art (SOA) and emerging issues concerning the biotechnical and soil bioengineering. Some of the relevant emerging applications are presented herein.

Biotechnology – Probably the hottest topic in erosion control today is "biotechnology", or the maximum use of vegetation. This technology appeals to environmentalists and engineers, alike, because it looks good and brings the many benefits of vegetation to bear on the problems of erosion and water quality. This is also why RECPs are steadily gaining in popularity. RECPs nurture and support vegetation, which, by itself, is often either, a fragile or a costly, labor-intensive construction material.

Anchored Geotextiles and Geonets – Gray and Sotir (1996) reveal that an anchored geonet presents a good example of an emerging development in biotechnical stabilization, specifically for stabilization of sandy slopes and coastal landforms. The anchored geotextile or geonet is made of a fabric or net which is stretched over and pulled down tightly on the ground surface and is secured in place by means of anchors inserted through and fastened to the geotextile.

Stabilization via Biological Organisms – An emerging technique for creating apparent cohesion within partially saturated sandy soil is via microbiological binding from the presence of fungal and bacterial colonies (Meadows et al. 1994). Incidentally, small quantity of cohesion or apparent cohesion significantly enhances the stability of sandy soils and slopes.

Other Emerging Technologies

Geosynthetics

Geotextile Tubes and Containers: Beneficial Uses – Very large geotextile containers filled with dredged material have re-gained popularity in the past few years because of their simple placement and construction, cost effectiveness and minimum impact on the environment. These containers are hydraulically or mechanically filled with a variety of dredged material types, including fine-grained materials.

Containment of dredged material in geotextile tubes, bags or other large containers, filled in place or filled in large bottom dump hopper barges and dumped below water has helped solve several difficult construction problems. Dike construction using long, sometimes continuous, tubes in wetlands, subdivision and perimeter dikes in dredged material disposal areas, under water stability berms, containment of contaminated materials, island construction, barrier island breach repair and structural scour protection are examples of projects that could not have been completed without use of geotextile containment systems. Sprague et al. (1998) has documented many large bag/tube/container projects constructed over the past 30+ years and reviewed common design and construction techniques.

Asphalt Reinforcement Layers – Countless miles of roads throughout the world must be maintained or improved annually to facilitate safe access to and commerce within and between populated areas. Over time, traffic volumes and wheel loads increase, accelerating the deterioration of road structures, which are already subject to the influences of climatic cycles, extreme weather events and lack of maintenance. This seemingly unavoidable deterioration has demanded cost-effective means for maintaining, i.e. rehabilitating and resurfacing, these facilities. One such cost-effective alternative has emerged which involves the incorporation of a geosynthetic within the bituminous resurfacing to produce a durable, weatherproof road surface that will resist reflective cracking and prolong pavement life.

Sprague et al. (1998b) noted that paving fabrics have been used for many years to retard reflective cracking and enhance waterproofing in asphalt overlays. More recently, higher strength geosynthetics have been incorporated into asphalt overlays to provide an even higher level of crack retardation and, in some cases, waterproofing. The mechanisms, which lead to the enhanced performance of reinforced overlays, have been described and quantified by Lytton (1989), demonstrating that the tensile stiffness added by the reinforcement allows the crack energy to be "intercepted" and reoriented horizontally.

Lightweight Fill, "Geofoam" – Although the use of geofoam as lightweight fill has been exploited in Scandinavia for many years, its application in U.S. is relatively recent. Due to its thermal insulation properties, it is primarily employed in conjunction with construction of highway fills over frost sensitive soils (Koerner 1998). Nevertheless, geofoam blocks can be utilized for construction over any type of soft soil (Horvath 1995).

Other recent (emerging) applications of geofoam are related to the environmental field (Koerner 1998) such as a spray for the daily landfill cover (Gasper 1990), to build a floating pond cover (Borgaard and Anderson 1994), to prevent the frost damage to clay liners, and to enhance the methane and radon gas venting (White 1995).

Summary and Future Trends

In this chapter, a brief overview of existing technologies for ground improvement is presented, along with emerging geomaterials for civil engineering applications. It is evident that during the past two decades in-situ ground modification has come of age, and numerous innovative techniques have been devised to resolve complex geotechnical engineering problems. An introduction to existing ground modification technologies, in the context of mechanical modification, hydraulic modification, physical and chemical modification, and modification by inclusions and confinement is presented in this chapter. Following a basic description, various types of geosynthetics including geotextiles, geogrids, geonets, geomembranes, geosynthetic clay liners (GCLs), geopipes, geofoams, and geocomposites are discussed with respect to geotechnical and geoenvironmental applications. Furthermore, other geomaterials such as soil admixtures, grouting, fiber-reinforced soils, soil nailing, soil bioengineering materials, prefabricated vertical drians, and erosion and sediment control materials are described briefly. Government regulations and standards of practice including USEPA's Subtitle C & D and AASHTO's geotextile and bridge specifications are outlined as well.

Emerging issues – durability, economics and proprietary restrictions are presented as a prelude to the state-of-the-art for emerging geomaterials and related technologies. The emerging geomaterials are described in the context of filtration and separation, stabilization and reinforcement, drainage, barrier, and erosion and sediment control technologies. Additional emerging technologies discussed here are geotextile tubes and containers, asphalt reinforcement layers, and lightweight fill.

The evolution of ground improvement involves the phases of application consideration, pilot-scale field testing, full-scale field testing, general field application, verification, engineering acceptance, and advanced field application (Welsh 1997b). While some of the emerging geomaterials described above have reached the stage of maturity, others are still at earlier stages of development and will require additional laboratory and field-testing and verification to become marketable. Nevertheless, the intent of developing and manufacturing these novel materials is to successfully meet the site improvement and remedial needs of engineering and construction community. Typical applications, concerning the civil engineering infrastructure, comprise increased bearing capacity, settlement control, soil stabilization, groundwater control, underpinning and excavation support, and liquefaction mitigation.

References

AASHTO (1997), "Interim Specifications - Standard Specification for Highway Bridges Division I, Section 5 – Retaining Walls & Division III, Section 7 - Earth Retaining Systems", American Association of State Highway and Transportation Officials (AASHTO), Washington, DC.

Agnew, W. and Theisen, M.S. (1994), " 'Built-in-place' Matrices, A Cost Effective Concept of Erosion Control", Proceedings of Conference XXV, IECA, Reno, NV, pp. 291-300.

Al-Alusi, H.R. (1997), "Compaction Grouting: From Practice to Theory," ASCE Geotechnical Special Publication, C. Vipulanandan, Editor, "Grouting: Compaction, Remediation and Testing," No. 66, pp. 43-54.

Anderson, S.A., and Hee, B.H. (1995), "Hydraulic Conductivity of Compacted Lateritic Soil with Bentonite Admixture," Environmental and Engineering Geoscience, Vol. I, No. 3, Fall, pp. 299-312.

"ASCE Preliminary Glossary of Terms Relating to Grouting" (1980), Committee on Grouting, Journal of Geotechnical Engineering Division, Vol. 106, No. GT7, pp.

803-815.

ASTM D4716 (1994), "Standard Guide for Testing Constant Head Hydraulic Transmissibility (In-Place Flow) of Geotextile and Geotextile Related Products."

Bandimere, S. W. (1997), "Compaction Grouting – State-of-the-Practice," Geotechnical Special Publications, Grouting: Compaction, Remediation and Testing, C. Vipulanandan, Editor, pp. 18-32.

Berg, R.R. (1993), "Guidelines for Design, Specification, and Contracting of Geosynthetic Mechanically Stabilized Earth Slopes on Firm Foundations, FHWA-SA-93-025, FHWA, Washington, D.C.

Bergado, D.T., Anderson, L.R., Miura, N., and Balasubramaniam, A.S. (1996), "Soft Ground Improvement in Lowland and other Environments, ASCE Press, New York, pp. 234-304.

Bogaard, D., and Anderson, R., "Geosynthetics Combine to Create an Efficient Floating Insulated Pond Cover," Geotechnical Fabrics Rpt., Vol. 12, No. 6, 1994, pp. 26-27.

Broms, B.B. (1993), "Lime Stabilization," Chapter 4, Ground Improvement, M.P. Moseley, Editor, Blackie Academic & Professional, Published by CRC Press, Inc., pp. 65-97.

Carroll, R.G., Jr., J.J. Rodencal and J.G. Collin (1992), "Geosynthetics in Erosion Control - The Principles", Geotextiles and Geomembranes, Vol. 11, pp. 523-534.

Chassie, R.G. (1994), "FHWA Ground Nailing Demonstration Project, Guideline Manual and Workshop," United States Federal Administration Publication.

Chesner, W.H., and Welsh, J.P., "The Use of Selected Waste Products for the Reclamation of Existing MSW Landfill Sites," Proceedings of National Waste Processing Conference, ASME, New York, NY, May 1992.

Coduto, D.P. (1999), Geotechnical Engineering, Principles and Practices, Prentice Hall, pp. 676-677.

Daniel, D. E. and Boardman, B. T. (1993) ``Report on Workshop on Geosynthetic Clay Liners", U.S. Environmental Protection Agency, EPA/600/R-93/171, August 1993.

Daniel, D. E. and Koerner, R. M. (1993) ``Technical Guidance Document: Quality Assurance and Quality Control for Waste Containment Facilities", U.S. Environmental Protection Agency, EPA/600/R-93/182, September 1993.

Eith, A. W., Boschuk, J., and Koerner, R. M. (1991), ``Prefabricated Bentonite Clay Liners", Geotextiles and Geomembranes, Vol. 10, Nos. 5-6, pp. 575-599.

Fang, H-Y. (1997), Introduction to Environmental Geotechnology, H-Y Fang, CRC Press, 1997.

Fang, H.Y. (1995), "Environmental Geotechnology," Chapter 2, Encyclopedia of Environmental Control Technology, Vol. 9, Geotechnical Applications, Leak Detection, and Treatment Options, P.N. Cheremisinoff, Editor, Gulf Publishing Co., Houston, TX, pp. 13-117.

Fang, H.Y. (1992), Topic 6, Geoenvironmental Engineering, Recent Accomplishments and Future Trends in Geomechanics in the 21st Century, U.S. – Canada Workshop, Univ. of Oklahoma and U.S. NSF, pp. 309-317

Foose, G.J., Benson, C.H., and Bosscher, P.J. (1996), "Sand Reinforced with Shredded Waste Tires," Journal of Geotechnical Engineering, ASCE, Vol. 122, No. 9, pp. 760-767.

Gasper, A.J., "Stabilized Foam as Landfill Daily Cover," Proceedings of Municipal Waste Management: Solutions for the 90's, Washington, D.C.: U.S.EPA, 1990, pp. 1113-1121.

Gray, D.H., and Al-Refeai, T. (1986), "Behavior of Fabric Versus Fiber Reinforced Sand," Journal of Geotechnical Engineering, ASCE, Vol. 112, No. 8, pp. 804-820.

Gray, D.H., and Ohashi, H. (1983), "Mechanics of Fiber Reinforcement in Sand," Journal of Geotechnical Engineering, ASCE, Vol. 112, No. 8, pp. 804-820.

Gray, D.H., and Sotir, R.B. (1996), Biotechnical and Soil Bioengineering Slope Stabilization – A Practical Guide for Erosion Control, Wiley Intersciences, pp. 337-359.

Gray, D.H., and Sotir, R.B. (1995), "Biotechnical Stabilization of Steepened Slopes," Transportation Research Record 1474, TRB, National Research Council, pp. 23-30.

Hansbo, S. (1993), "Band Drains," Chapter 3 of "Ground Improvement," Edited by M.P. Moseley, Blackie Academic & Professional (Chapman & Hall), 1993.

Hausmann, M.R. (1990), Engineering Principles of Ground Modification, McGraw-Hill Publishing Co.

Head, W.J., McQuade, P.V., and Anderson, R.B. (1982), "Coal Refuse and Fly Ash Compositions: Potential Highway Base Course Materials, Transportation Research Record No. 839, TRB, National Research Council, Washington, D.C., pp. 11-19.

Horvath, J.S. (1994), "Expanded Polystyrene (EPS) Geofoam: An Introduction to Material Behavior," Journal of Geotextiles and Geomembranes, Vol. 13, No.4, pp. 263-280.

Horvath, J.S., ed., Geofoam Geosynthetic, Scarsdale, NY: Horvath Engineering, P.C., Scarsdale, NY, 1995.

Jewell, R.A. (1980), "Some Factors Which Influence the Shear Strength of Reinforced Sand," Proceedings, Symposium on Common Applications to Geotechnical Problems in Highway Engineering, CUED/D-Soils/TR85.

Kauschinger, J.L., Perry, E.B., and Hankour, R. (1992), "Jet Grouting: State-of-the-Practice," Geotechnical Special Publication: Grouting, Soil Improvement and Geosynthetics, Vol. 1, No. 30, pp. 169-182.

Khan, L.I., and Sarker, M. (1993), "Enzyme Enhanced Stabilization of Soil and Fly Ash," ASCE Geotechnical Special Publication: Fly Ash for Soil Improvement, No. 36, pp. 43-57.

Koerner, R. M. (1998), Designing with Geosynthetics, 4th Edition, Prentice-Hall, New Jersey.

Koerner, R. M. (1991), "Geosynthetics in Geotechnical Engineering," Chapter 22, Foundation Engineering Handbook, H-Y Fang, Editor, 2nd Ed., pp. 796-813.

Koerner, R.M., and Hwu, B-L. (1990), "Prefabricated Highway Edge Drains," ASCE/PennDot Geotechnical Seminar, Hershey, PA.

Koutsourais, M. and Sprague, J. (1992), "An Inside Look at Articulating Concrete Block Revetments", Proceedings of Conference XXIII, IECA, Reno, NV, pp. 49-55.

Lancaster, T. and Austin, D.N. (1994), "Classifying Rolled Erosion Control Products: A Current Perspective", Geotechnical Fabrics Report, Oct/Nov, pp. 16-22.

Leflaive, E. (1985), " Soils Reinforced with Continuous Yarns: the Texol," Proceedings of 11th International Conference on Soil Mechanics and Foundation Engineering, San Francisco, Vol. 3, pp. 1787-1790.

Maher, M.H., Butziger, J.M., DiSalvo, D.L., and Oweis, I.S. (1993), " Lime Sludge Amended Fly Ash for Utilization as an Engineering Material," ASCE

Geotechnical Special Publication: Fly Ash for Soil Improvement, No. 36, pp. 73-88.

Maher, M.H., and Gray, D.H. (1990), "Static Response of Sand Reinforced with Randomly Distributed Fibers," Journal of Geotechnical Engineering, ASCE, Vol. 116, No. 11, pp. 1661-1677.

Maher, M.H., and Woods, R.D. (1990), "Dynamic Response of Sand Reinforced with Randomly Distributed Fibers," Journal of Geotechnical Engineering, ASCE, Vol. 116, No. 7, pp. 1116-1131.

McGown, A. et al. (1985), "Soil Strengthening Using Randomly Distributed Mesh Elements," Proceedings of 11th International Conference on Soil Mechanics and Foundation Engineering, San Francisco, Vol. 3, pp. 1735-1738.

Meadows, A., Meadows, P.S., Wood, D.M., and Murray, J.H.M. (1994), Microbiological Effects on Slope Stability: An Experimental Analysis," Sedimentology 41.

Mitchell, J.K., Baxter, C.D.P., and Munson, T.C. (1995), "Performance of Improved Ground During Earthquakes," in Soil Improvement for Earthquake Mitigation, Edited by R.D. Hryciw, Geotechnical Special Publication No. 49, American Society of Civil Engineers, New York, NY, pp. 1-36.

Mitchell, J.K., and Zornberg, J.G. (1995), "Reinforced Soil Structures with Poorly Draining Backfills, Part II: Case Histories and Applications", Geosynthetics International, Vol. 2, No. 1, pp. 265-307.

Mitchell, J.K., and Zornberg, J.G. (1994), "Reinforced Soil Structures with Poorly Draining Backfills, Part I: Reinforcement Interactions and Functions", Geosynthetics International, Vol. 1, No. 2, pp. 103-147.

Mitchell, J.K., and Van Court, W.A. (1992), "The Role of Soil Modification in Environmental Engineering Applications," Geotechnical Special Publication: Grouting, Soil Improvement and Geosynthetics, Vol. 1, No. 30, pp. 110-144.

Moseley, M.P., Editor (1993), Ground Improvement, Blackie Academic & Professional (An Imprint of Chapman & Hall), Published in USA by CRC Press, Inc.

Pearlman, S. (1998), "Jet Grouting: New Directions," Civil Engineering, Vol. 68, No. 8, pp. 40-44, Aug. 98.

Schaefer, V.R., Editor (1997), "Ground Improvement, Ground Reinforcement, and Ground Treatment Developments 1987-97," Geotechnical Special Publication No.69, American Society of Civil Engineers (ASCE).

Shea, B.M. (1993), “Tips to Succeeding with Silt Fence,” Geotechnical Fabric Report (GFR), Vol. 14, No. 7, pp. 30-33, Sept. 93.

Sprague, C.J. (1998-submitted), "Geosynthetic Reinforcement: Are Geotextiles and Geogrids Interchangeable", Proc. of the 6th Int'l Conf. on Geotextiles, Geomembranes and Related Products, Atlanta, pp.

Sprague, C.J., Bradley, A.S., Toups, D., and Trainer, E. (1998-submitted), "Geotextile Tubes - Case Histories and Lessons Learned", Proceedings of Conference XXIX, Int'l Erosion Control Assn., Reno, NV, pp.

Sprague, C.J., Allen, S., and W. Tribbett (1998-submitted), "Tensile Properties of Asphalt Overlay Geosynthetic Reinforcement”, Transportation Research Board, Washington, DC.

Sprague, C. J. and Goodrum, R. A. (1994), ``Selecting Standard Test Methods for the Experimental Evaluation of Geosynthetic Durability", Transportation Research Record No. 1439, TRB, Washington, DC, pp. 32-40.

Sprague, C.J. and Koutsourais, M.M. (1992), "Fabric Formed Concrete Revetment Systems", Geotextiles and Geomembranes, Vol. 11, pp. 587-609.

Tarricone, P. (1994), “Jet Grouting Gains,” Civil Engineering, Vol. 64, No. 12, pp. 40-43, Dec. 94.

Theisen, M.S. (1992), "Evaluation of Biotechnical Composites Under High Velocity and Shear Conditions", Proceedings of Conference XXIII, IECA, Reno, NV, pp. 285-305.

Theisen, M.S. (1993), “The Expanding Role of Geosynthetics in Erosion and Sediment Control,” Land and Water, Jan/Feb, pp. 30-35.

Theisen, M.S. and Carroll, R.G. (1990), "Turf Reinforcement - The 'Soft Armor' Alternative", Proceedings of Conference XXI, IECA, Washington, D.C., pp. 257-270.

Trotti, J. (1994), "Wildfire: A Challenge for Erosion Control", Erosion Control, March/April, pp. 34-43.

Vipulanandan, C., Editor (1997), “Grouting: Compaction, Remediation and Testing,” ASCE Geotechnical Special Publication, No. 66.

Welsh, J.P. (1997a), “State-of-the-Art of Grouting in North America,” Proceedings of IS-Tokyo ’96 / The 2nd International Conference on Ground Improvement Geosystems, Tokyo, Japan, May 14-17, pp. 825-831.

Welsh, J.P. (1997b), "Excellence in Geotechnical Engineering Ground Modification Milestones," Proceedings of the Central Pennsylvania Section, ASCE and PennDOT, Harrisburg, PA.

Welsh, J.P. (1992), "In-Situ Ground Modification for New and Remediation Construction," Proceedings of Ohio River Valley Soil Seminar (ORVSS XXIII), Louisville, KY.

Welsh, J.P., Editor, (1987), "Soil Improvement – A Ten Year Update," Proceedings of a Symposium Sponsored by the Committee on Placement and Improvement of Soils, Geotechnical Engineering Division, ASCE, Geotechnical Special Publication, No. 12.

West, H. (1994), "The Function of Continuous Cover in Preventing Soil Erosion from Rainfall Impact and Sheet Flow: The Evolution of Bonded Fiber Matrices," Proceedings of Conference XXV, Reno, NV, pp. 349-354.

White, R., "EPS Used to Assist in Methane and Radon Gas Venting, "Geosynthetics World, Vol. 5, No.2, 1995, p.12.

Winterkorn, H.F., and Pamukcu, S. (1991), "Soil Stabilization and Grouting," Chapter 9, Foundation Engineering Handbook, H-Y Fang, Editor, 2nd Ed., pp. 317-369.

Wu, K.J. and Austin, D.N. (1992), "Three-Dimensional Polyethylene Geocells in Erosion Control and Channel Linings", Geotextiles and Geomembranes, Vol. 11, pp. 611-620.

Zoghi, M., and Bodocsi, A. (1995), "Emerging Technologies in Ground Modifications for Environmental Control," Chapter 1, Encyclopedia of Environmental Control Technology, Vol. 9, Geotechnical Applications, Leak Detection, and Treatment Options, P.N. Cheremisinoff, Editor, Gulf Publishing Co., Houston, TX. pp 1-13.

Craig C. Menzemer[1]

Chapter 4

Aluminum Materials and the Infrastructure

Abstract

Continued increase in the use of aluminum in the infrastructure is likely, given the physical characteristics, flexibility of product forms, relative ease of fabrication, intelligent application and continued material development. While often market specific, new materials usually find more than a single use. Efforts to reduce energy content, minimization of process byproducts and development of thin gage sheet/plate casting over a wide range of alloys should help to lower initial costs. Further, maturation of life cycle cost methodologies will benefit utilization of aluminum as long term maintenance is generally reduced.

Physical characteristics beneficial to infrastructure application are reviewed, as well as recent efforts to develop new aluminum alloys. Composition and mechanical properties are examined, along with some recent applications of new materials and new uses for traditional alloys. Research needs and the future outlook are summarized.

Keywords: aluminum, infrastructure, fabrication, joining, bridges, transportation

Introduction

Many recent alloy developments in aluminum are application specific and as such, are only peripherally related to traditional infrastructure systems. Although targeted for specific markets, many of the newer alloys should find application elsewhere. Some newer constructional alloys, on the other hand, are considered to be variants on traditional material systems. Initial modification began with simple composition and treatment modification, and in several instances, resulted in significant physical property changes.

In general, infrastructure type applications for aluminum alloys have been limited to secondary structures like light poles, overhead sign trusses, small culverts,

[1] Dr. Craig C. Menzemer, M. ASCE, Assistant Professor, Department of Civil Engineering, The University of Akron, Akron, Ohio 44325-3905

building facades, roof coverings, storage tanks, etc... In only a few instances have traditional aluminum alloys and product forms been employed and universally accepted as primary load carrying elements within an infrastructure system despite technical success in different applications. One example would be lightweight replacement Bridge decking (Menzemer et al. 1987). Barriers to implementation include a lack of familiarity with design rules, fabrication procedures and initial costs.

Aluminum offers properties attractive to those that design, fabricate and operate infrastructure systems. Many constructional alloys possess high strength to weight ratios. Aluminum may be substituted for traditional materials with a decrease in weight and no appreciable loss in strength. Weight reduction provides for ease of field erection and shop fabrication of larger components and subassemblies allows for enhanced quality control. In addition, corrosion resistance for constructional alloys subjected to typical atmospheric exposures is exceptional, and often eliminates painting. As a result, maintenance costs are reduced.

Wrought and cast products are available in a number of forms. Wrought products include sheet, plate, extrusion and forging. While many standard cross sections are available, the extrusion process provides the capability for customizing shapes directly from the mill. This may eliminate fabrication process steps associated with the build up of a cross section from plates, as often required for other materials. It also provides the designer the ability to place material where needed and the opportunity to reduce wall thickness in other, more lightly stressed areas. An example of optimal material placement might be the addition of a stiffening lip to increase buckling resistance. Other features may be incorporated into the extrusion as well, including flanges for gaskets, etc... Extrusion size is limited by press capacity which, varies from manufacturer to manufacturer. Extrusions encompassing a 30 in circle size are possible, but members fitting a 15 in circle size are more common.

Casting process technology has advanced rapidly over the past several years. Capability to produce relatively large, thin wall (2mm) finished parts is a reality using vacuum assisted die-casting. Alloy refinement, casting equipment, operating procedures and metal handling processes have all played a role. Near finished parts may be produced with strengths reaching close to 30 ksi and elongations in excess of 10%. Standard casting processes have advanced to the point where mechanical properties are tailored to regions within a component through an integrated design – manufacturing framework (Hinkle et al. 1996). As with the extrusion process, castings may be attractive for both design and fabrication. Castings provide the ability to place metal where needed, while the near net shape eliminates piece count associated with the fabrication of parts from stamped sheet metal components or cut plate.

Semi-Solid-Forming (SSF) has recently been developed into production robustness for aluminum (Zehe 1997). A majority of the alloys employed fall within operating windows of currently registered materials. Originally referred to as the thixocast process, SSF is considered a cross between casting and forging. SSF begins with a billet containing a globular, rather than a dendritic type structure. The globular structure is developed as a result of metal agitation during billet solidification. These materials will behave as a solid until subjected to sufficient shear forces present in the forming operation (Boylan 1997). Upon heating and subsequent forming, there is a change in apparent viscosity allowing the metal to act like a viscous liquid. Flow becomes much more uniform, eliminating some of the difficulties associated with lubricant and air entrapment. Minimization of solidification shrinkage and lowering total heat content tends to elevate mechanical properties. Once formed, the parts undergo quenching and subsequent heat treatment. Both microstructure refinement and property improvements result.

Being a cross between casting and forging, SSF develops near net shaped parts without other significant fabrication processes. Complex shapes are possible, allowing the designer to place material where needed. Improved properties enable use in structural applications. Currently, the majority of use is geared toward the transportation industry, with development of suspension components, valve bodies, master cylinders, etc. Other advantages include reduction of part count in complex components.

At the point where a structure or system becomes no longer useful or obsolete, aluminum offers the advantage of being recyclable. The residual value of aluminum in the system may be significant, and should be considered as part of any life cycle cost analysis. In the automotive industry alone, about 85-90% of the aluminum from scrap automobiles is recycled (Aluminum Association 1995). Recycling of aluminum not only conserves resources used during reduction and extraction of primary metal, but reduces the need to dispose of wastes generated as a result of the smelting process. Airborne emissions are drastically reduced as well. Nearly one-third of the total aluminum supply is recycled, and represents an important source of primary metal.

A key to recycling aluminum is the flexibility of alloy groups. All alloys, which are of compatible composition, may be handled together. With the majority of structural components fabricated from 3XXX, 5XXX and 6XXX series alloys, recycling of components should be a natural consequence of the life cycle. Ideally, those utilizing aluminum alloys in infrastructure design should consider recycling early on in the process. Specifying similar alloys and alloy groups for components that are difficult to separate enhances residual monetary value through direct labor reduction.

Metallurgical considerations

In the pure or unalloyed state, aluminum is a soft, ductile metal, which exhibits good electrical and thermal conductivity. In the pure form, it has a high resistance to atmospheric corrosion. A major drawback in the unalloyed state is the greatly reduced strength. Alloy 1060 is 99.6% aluminum, and exhibits a typical yield strength of 4 ksi, an ultimate of 10 ksi and 43% elongation (Aluminum Association 1988).

Aluminum gains strength by alloying in combination with thermal heat treatment or cold work. Alloying elements that form a solid solution with aluminum are copper, manganese, magnesium, silicon and zinc. Magnesium, zinc and copper provide strengthening. Magnesium enhances corrosion resistance as well. Silicon increases fluidity at higher temperatures and as such, is found in many common extrusion alloys as well as welding wire. Manganese provides for increased ductility as well as some modest strengthening.

Alloying depends on the ability of the element to form a solution or mixture with aluminum as a solid. Aluminum will accept various amounts of the principal alloying elements. Zinc shows the maximum solid solubility followed by magnesium. Of any single primary alloying element, silicon shows the smallest solid solubility at about 1.6% (Aluminum Association 1991).

The particular elements used in the alloy will determine whether or not the material is heat-treatable or non-heat-treatable. Materials employing magnesium in combination with silicon or zinc and those utilizing copper as primary elements are heat-treatable. When magnesium, or manganese or silicon is used as a primary alloying element, the material is considered to be non-heat-treatable. Strength enhancements are then gained through cold work. Secondary elements are sometimes added to enhance particular properties. Included are heat treat response, corrosion resistance and strength. These secondary elements are usually not considered when examining the basic material response.

Joining processes

Aluminum may be joined by any standard processes associated with other metals fabrication, including mechanical fasteners, welding, adhesive bonding, or by a combination techniques. Mechanical fasteners may be rivets, bolts or a form of mechanical clinch. The rivet or bolt material may be aluminum or a coated steel product. Coating of steel fasteners is generally necessary to prevent galvanic corrosion.

Welding is one of the most common techniques used to join aluminum components. Standard procedures include a variety of arc and resistance processes. Laser, electron beam, electroslag, electrogas and oxygas welding are some other techniques used on a smaller scale. A newer, autogenous process is known as friction stir welding (Edison Welding Institute 1996, ESAB 1997). Friction stir welding uses tooling similar to an end mill that tracks along the joint seam. A pin is mounted in a chuck that rotates at a relatively high speed. After piercing a hole at the beginning of the joint, the pin is moved forward along the seam. Friction between the pin and base metal creates heat sufficient to plasticize material adjacent to the pin. Material in back of the pin is forced together to form a bond. Since no melting of the base material occurs, metal along the fusion line is left in a fine-grained, but hot worked condition. Higher joint efficiencies have been reported. One primary disadvantage includes the relatively large initial equipment costs. Further, the size of the equipment limits use to shop fabrication and has not yet been developed sufficiently for field applications.

Adhesive bonding has been studied as a primary joining method for aluminum components and structures. While surface preparation added significant steps to early processes, newer adhesives have been developed which require little in the way of uniquely clean conditions. Design methodologies have been developed and refined as well (Sharp et al. 1996; Hart-Smith 1987). Difficulties with joint inspection and questions regarding long term durability remain. As such, joints that utilize adhesives will often be employed in conjunction with mechanical fasteners.

New Alloys for Infrastructure Use

Many recent alloy development programs have been targeted for specific applications, and have been constrained by narrow functional requirements. While not all are infrastructure related, the alloys should find application in areas other than those prompting initial development. This is particularly true of the 5XXX and 6XXX series materials developed to provide additional design features like enhanced base and as-welded strengths, corrosion resistance and formability. Other classes of new materials include a weldable variant of 7108 that is primarily used in structural shapes. Further, aluminum metal matrix composites are finding increasing applications.

Two primary classes of wrought alloys for general construction are the 5XXX and 6XXX series. 5XXX series materials utilize magnesium as the primary alloying element, and are non-heat-treatable. Non-heat-treatable alloys gain strength through cold work. The amount of cold work is controlled so as to produce the desired properties. Cold work is achieved by stretching some alloys and rolling others. Materials that have been cold worked show a gain in tensile yield strength accompanied by a smaller increase in ultimate. Ductility is reduced with increasing amounts of strain hardening. An important distinction among 5XXX series aluminum alloys is the percentage of magnesium in solution. Alloys may be classified into two groups, those containing less than 3% magnesium and those containing more than 3% magnesium. Higher percentages of magnesium increase strength of the material. However, resistance to stress corrosion cracking and exfoliation are diminished. Alloys possessing greater than 3% magnesium are not well suited for elevated temperature applications.

Among two new 5XXX series alloys are 5383 and a material not currently registered, but is referred to as "Alustar". Both alloys were developed for the marine industry , and introduced as an alternative to the current version of 5083 (Raynaud et al. 1997; Sampath et al. 1997). Tempers being targeted are H116/H321. Aluminum alloy 5383 was first envisioned as an optimized 5083, but composition changes coupled with processing improvements resulted in substantial property enhancements. Alustar development was driven by the need to reduce weight of high speed craft hull structures. In both instances, increased parent metal and as welded strength compared to 5083 was desired, but with no sacrifice in corrosion resistance, weldability or formability.

Table 1 provides the composition limits for 5383. In general, 5383 has reduced iron and silicon levels, but increased magnesium, manganese, copper, zinc and zirconium contents as compared to 5083. Increased levels of Mg, Zn, and Cu would be expected to enhance strength, while additional Mn would tend to increase ductility. Table 2 compares typical tensile properties for both base and welded test coupons against 5083. While the increase in yield strength for a wide range of product thickness is modest for base material, an improvement in both ultimate strength and elongation are noted. Welded yield strengths are improved by nearly 15%. Although greater than 4% Mg, 5383 demonstrates resistance to exfoliation and pitting corrosion. Application of 5383 is expected to result in decreased weight of components and structures, particularly in areas governed by static strength. Fatigue behavior is not likely to be affected by simple composition and treatment changes.

Table 1 – *Composition Limits for Aluminum Alloy 5383*

Element	Minimum (%)	Maximum (%)
Si	---	0.25
Fe	---	0.25
Cu	---	0.20
Mn	0.70	1.0
Mg	4.0	5.2
Cr	---	0.25
Zn	---	0.40
Ti	---	0.15
Zr	---	0.2

Table 2 – *Typical Tensile Properties of Base and Welded 5383 and 5083*

Alloy/Property	5083-H116	5383-H116
Yield Strength (MPa)	257	257
Ultimate Strength (MPa)	351	377
Elongation (%)	18.7	22.2
Welded Yield (MPa)	134	149
Welded Ultimate (MPa)	287	299
Welded Elongation (%)	12.8	10.2

"Alustar" is another 5XXX series alloy developed for the marine industry. Again, the goal was to increase strength in both the base and as-welded strength conditions. A 20% increase in both the yield and ultimate parent metal strengths has been reported as compared to 5083. Annealed strength is also improved, as evidenced by tensile properties developed on as-welded coupons. Such improvements have been attributed to a reduction in Fe and Si content, while increasing the level of Zn.

Corrosion characteristics of the aluminum magnesium alloys are dependent upon magnesium content, grain structure (fabrication history) and environment. AlMg particles that do not remain in solid solution are anodic as compared to the remaining aluminum matrix. Under some conditions, these particles will congregate along grain boundaries or along deformation induced slip planes (Hatch 1984). Corrosion of the anodic particles would lead to selective attack along preferential directions. Corrosion resistance is provided in the higher magnesium content aluminum alloys by fabrication procedures that ensure random dispersion of the AlMg constituent particles. Both 5383 and Alustar develop corrosion resistance suitable for marine use, as does 5083.

One of the newer 6XXX series alloys has been developed for use in forged and extruded components. As with the 5XXX materials discussed, 6069-T6 develops tensile yield and ultimate strengths in excess of 6061-T6 (Bergsma et al. 1996). Table 3 compares typical tensile properties and Table 4 compares composition windows for 6061 and 6069. As may be seen, 6069 has increased silicon, copper and magnesium as compared to 6061. These additions would be expected to increase tensile properties. Not unlike some of the higher strength 6XXX series alloys, 6069 contains silicon in excess of the solid solubility limit. However, the magnesium content has been increased as well. Despite composition changes, sustained load cracking resistance appears to be equivalent to 6061-T6 as does corrosion resistance. Other benefits include improved surface conditions and scrap compatibility with many of the 5XXX and 6XXX alloys.

Table 3 – *Typical Tensile Properties of 6061-T6 and 6069-T6*

Alloy/Property	6061-T6 (Extruded Bar)	6069-T6 (Extruded Shape)
Yield Strength (MPa)	275	346
Ultimate Strength (MPa)	310	396
Elongation (%)	12	20.9

Table 4 – *Composition Limits for 6061 and 6069*

Element	6069 (%)	6061 (%)
Si	0.6-1.2	0.4-0.8
Fe	0.4	0.7
Cu	0.55-1.0	0.15-0.4
Mn	0.05	0.15
Mg	1.2-1.6	0.8-1.2
Cr	0.05-0.3	0.04-0.35
Zn	0.05	0.25
Ti	0.1	0.15
V	0.10-0.30	---

Modification of 7108, or 7108.50, has resulted in a weldable, high strength aluminum alloy for application in the transportation industry (Hval 1997). Narrow composition control combined with a heat treatment to a slightly over-aged condition has provided higher base metal and as-welded strengths as compared to traditional 6XXX extrusion alloys. Table 5 provides the composition window. While stress corrosion resistance has been a concern for some 7XXX series materials, over-aging combined with zirconium additions designed to minimize recrystallization during extrusion and welding have appeared to reduce the alloy's susceptibility.

Table 5 – *Composition Limits for 7108.5*

Element	Minimum (%)	Maximum (%)
Zn	5.2	5.7
Mg	0.7	1.0
Zr	0.12	0.25
Cu	---	0.05

Table 6 compares tensile strength of the modified 7108.05-T79 with 6061-T6 in the parent metal condition. As may be seen, 7108.50 develops higher tensile properties. In addition, the new alloy has the ability to recover strength in the heat-affected-zone subsequent to welding through natural aging. Aging occurs during a one to four week period after welding, and develops during storage at room temperature. Dilution of acceptable filler alloys with the base material results in a combination of zinc and magnesium, similar to that of the parent material. In effect, the weldment becomes "heat-treatable". In addition, sensitivity of the new alloy to hot-short cracking from overheating during welding is much less than traditional 6XXX series materials.

Table 6 – *Minimum Tensile Properties of 7108.5-T79 and 6061-T6*

Property	7108.5-T79	6061-T6
Yield Strength (MPa)	290	242
Ultimate Strength (MPa)	330	290
Elongation (%)	10	12

Other recently developed aluminum based materials include a series of discontinuously reinforced metal matrix composites (MMC). Several of these utilize silicon-carbide ceramic particles as reinforcement (DWA 1997; DURALCAN 1996). Both cast and wrought product forms are available that include gravity and die castings, extrusion and sheet products. Different processes have been developed for fabrication of finished mill goods. While the lightweight features of aluminum are retained, there is a significant increase in elastic modulus, up to 40% as compared to the standard family of alloys. As with traditional aluminum materials, a range of tensile properties is available. Table 7 summarizes some of the tensile properties for commercial aluminum MMC. As may be noted, some of the MMC materials exhibit strengths close to that of aerospace alloys, although elongation values tend to be lower. Fatigue crack growth characteristics are improved at low levels of crack driving force as compared to 6061-T6.

Joining aluminum MMC components has been achieved using a variety of standard techniques. Among these are welding and mechanical fastening. Riveted and bolted joints take advantage of the higher bearing strength afforded by the MMC materials. Welding is accomplished with standard filler alloys.

Table 7 – *Minimum Tensile Properties of Some MMC Materials*

Property	6092/SiC/25p	6092/SiC/17.5p	W6A.15A-T6	W6A.20A-T6
Yield (MPa)	449	421	324	317
Ultimate (MPa)	538	490	338	345
Elongation (%)	4	6	6	4

Bridge Applications

Although not universally accepted, aluminum alloys and products have been used as primary load carrying members in bridges. Approximately ten bridges have been built out of aluminum since 1933, and each application considered a technical success (Aluminum Association 1996). Attractive properties include a relatively high strength to weight ratio, with minimum guaranteed properties approaching those of mild structural grades of steel. Further, the ductility and toughness of most aluminum alloys increase as the temperature decreases, reducing concerns surrounding transition temperatures and the possibility of brittle fracture. A majority of the alloys employed in infrastructure systems are corrosion resistant, eliminating the need for periodic painting, thus lowering maintenance costs.

Standard bridge fabrication practices favor aluminum with increased machining, cutting and drilling speeds. Load levels required for punching are reduced as well. Cutting may employ techniques similar to those used for other metals. However, plasma arc cutting of heat treatable alloys has resulted in hot - short cracks from the melting of low temperature constituents along grain boundaries. If used for alloys that gain strength by aging or heat treatment, mechanical removal of material from along the cut edge is necessary.

Recent efforts have been concentrated in modular replacement decking, with several projects completed or underway. Comparable components are lightweight and easy to maneuver and handle in both the shop and field. Bridge decks designed from aluminum have been a stiffened plate or multi-void cell type structure, with weights ranging from 15 to 25 psf. Typical reinforced concrete decks may weigh up to 100 psf, with steel counterparts approximately twice that of the aluminum systems. As such, allowable load ratings may be increased through deck replacement with minimal retrofit of the superstructure or substructure. Drastic dead weight reduction may, in fact, salvage some bridges slated for replacement. Modular in nature, individual panels are shop fabricated to the point that a lightweight wearing surface is applied. Typically, these panels may be 10 ft wide and up to 40 ft in length. Prefabricated panel construction may reduce overall costs as portions of the roadway may be closed during off peak traffic hours to allow for demolition and subsequent deck replacement.

While reduction in dead weight benefits the structure in terms of retrofit and repair costs, subsequent increases in live load rating are not always warranted (Kosteas et al. 1995). Fatigue resistance of geometrically similar steel and aluminum details vary, with ratios ranging from 1/3 to 1/(2.5). Attention to section proportioning and use of good practice detailing is critical to success of aluminum modular decking systems. Many of the systems employ bolted connections as well as groove-welded details.

Figures 1 to 3 illustrate several systems currently being developed and employed on bridges (Aluminum Association 1996). Figure 1 shows an orthotropic deck that consists of several different extruded shapes welded to a deck plate. An epoxy-wearing surface is applied to the deck, and is nominally 3/8 in thick. Galvanic isolation from the steel superstructure is achieved through stainless steel shims and polymer bearing pads. Hold-down extrusions are employed on adjacent sections to allow for thermal expansion and contraction. Figure 2 depicts the multi-void cellular structure. As before, the wearing surface is shop applied and consists of a two-part epoxy with abrasive material embedded in the adhesive. Each deck section is built up from a series of extruded shapes, being joined by groove welds. Galvanic isolation is provided by effective separation of the superstructure from the deck. Figure 3 depicts another system currently being used in Sweden. It is a multi-void cellular deck built out of extruded profiles. The deck employs tongue and groove joints between panels, which are bolted down to the superstructure with stainless steel clamps.

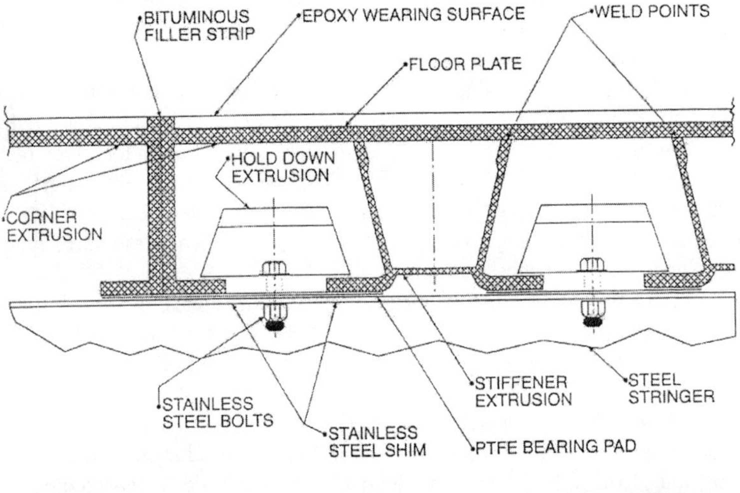

Figure 1 – *Example of an Aluminum Orthotropic Bridge Deck*

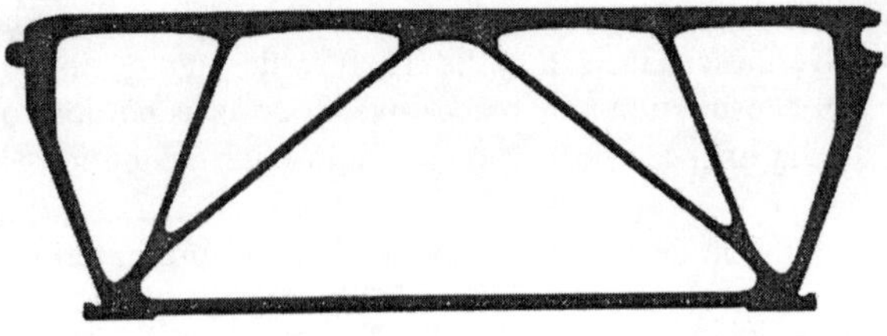

Figure 2 – *An Example of a Multi-Void Hollow Bridge Deck Section*

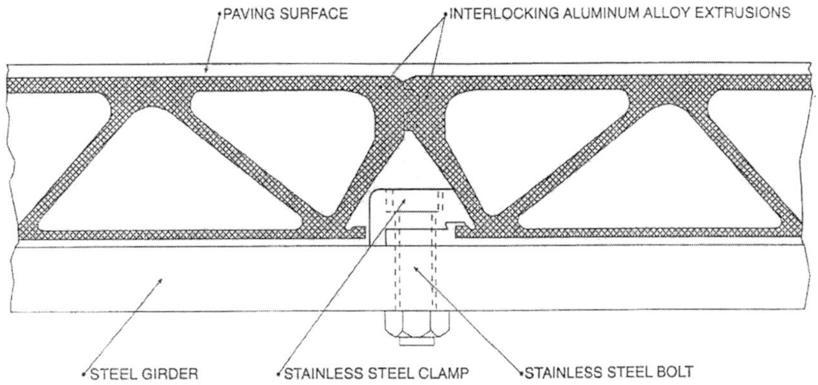

Figure 3 – *Example of an Aluminum Bridge Deck Section Utilized in Sweden*

A recent example of an all aluminum structure for highway traffic is the Forsmo Bridge in Norway (Kvale et al. 1997). At a length of 128 ft, continuous over two equal spans, it is an example of an all welded design that was built entirely in the shop. Project goals included minimizing stress concentrations, structural optimization based on weight and design for ease of manufacture. Design and fabrication took place over just an 18-week period of time. Once completed, the structure was transported to the construction site by a freighter and subsequently by truck. Two mobile cranes were used to perform the lift and place the nearly 31 ton bridge on the foundations. Total erection time was less than an hour, with the bridge being opened to traffic the following day.

A majority of the alloys used in construction were 6XXX series extruded profiles. A smaller tonnage of 5083 plate was utilized around the points of support. The structure is a twin trapezoidal box girder with an orthotropic deck as the top flange. Each box section was built from open extrusions with integrated longitudinal stiffeners. Internal cross bracing placed every 9.8 ft serves to minimize distortion of the boxes and assist with lateral load distribution. The 5083 plate serves as webs and bottom flanges of the box sections at the foundations. In addition, built up cross sections were used for lateral cross bracing. A schematic cross section of the structure is shown in Figure 4.

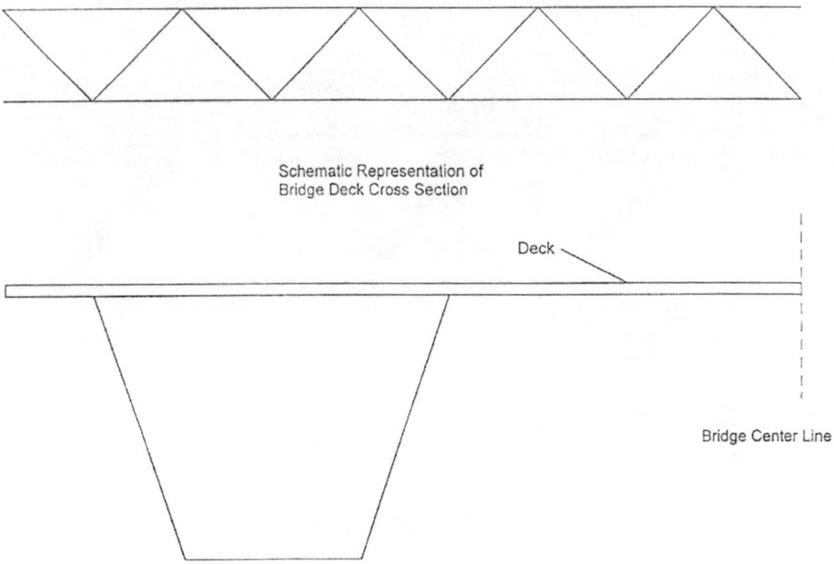

Figure 4 – *Schematic Representation of Forsmo Bridge Cross Section*

Fabrication of the box girders began with the development of web and bottom flange panels. Panels were fabricated in 39.4 ft lengths utilizing fabrication fixtures. One of the principal challenges in the design of fixtures, setting of weld parameters and sequencing is the minimization of distortion. Welding was performed using the gas metal arc process, but with an inverter type power supply. Joining webs to bottom flange, and installation of cross bracing completed box sections. The orthotropic deck was placed as the top flange by welding in the flat position.

The orthotropic deck is welded to each trapezoidal box, and effectively serves as the top flange. A multi-hollow extruded profile is used for the deck, with each profile joined by longitudinal groove welds. Asphalt is used as the wearing surface. In order to achieve a sufficient bond with the aluminum deck, the top surface was grit blasted, painted with an epoxy and sprinkled with dry sand. Asphalt was applied in the field using standard installation methods.

Transportation

While the term "infrastructure" implies a fixed state or permanence, a major market for increased use of aluminum is transportation. Although increasing amounts of aluminum in the automobile have been fairly well documented, the transport of bulk commodities has resulted in increased demand for aluminum railcars, trailers, and containers. These products should be considered as a major part or perhaps a customer of the infrastructure. Most applications may be considered a new use for more traditional materials.

Aluminum has been used in the building of railcars for a number of years, although primarily limited to coal cars. Recent efforts have focused on the development of covered hopper cars for movement of bulk commodities. Attributes of aluminum beneficial to the railroads include lightweight, high specific strength, ease of fabrication and low maintenance. As may be obvious, aluminum's lightweight and high specific strength translates into economic benefits for railroad operations. This is particularly true for highly mobile operations like unit train service. Typically, a 20% car tare weight saving is realized over cars built of traditional metals (Progressive 1986, Aluminum Association 1980). Given a fixed weight on rails for freight service, additional product may be moved each trip. Increased payload reduces the number of trips required; thereby reducing both fuel and labor costs.

Maintenance is an issue with many current covered hopper cars constructed from structural grades of steel. These cars require a polymer lining so as to isolate the load from the car and to avoid contamination from scale and rust. These lining materials have to be replaced several times over the life of the car. In addition, steel cars often require exterior painting. Linings and paint increase maintenance costs over a comparable aluminum railcar, and may be used as justification to offset an initial cost differential.

Figures 5 and 6 illustrate some differences in the use of aluminum in railcars. Figure 5 shows a traditional design scheme applied to an aluminum car. A continuous steel center sill and truck arrangement is used, with an aluminum box constructed on top of the substructure. Top and bottom chords are extruded 6061-T6 aluminum sections, mechanically fastened to 5083-side plates. Each plate is reinforced with several extruded hat-shaped sections designed to resist shear from loads transferred around the bolster area of the car.

Figure 5 – *Aluminum Covered Hopper Car*

Figure 6 illustrates a relatively new car designed by taking advantage of the specific attributes of aluminum. The car utilizes a stub sill that provides for additional payload volume, but creates a challenge in the transfer of axial load from the stub center sill to the side sill members. Both top and bottom chords are 6061-T6 with side plates constructed from 5083-H321. In this case, additional buckling resistance in the side is developed through the curved geometry rather than application of vertical stiffeners. Curved geometry provides for an increase in volume as well. A vast majority of the joints are welded. This limits objects like fastener heads that might protrude into the interior of the car. Product flow during unloading is improved, and substantially reduces the incidence of product hang-up.

Figure 6 – *Aluminum Covered Hopper Car*

Another example of the use of aluminum in the transportation of bulk commodities is shown in Figure 7. An intermodal shipping container may be loaded with product and transported by rail, tractor-trailer or boat. Frame members and stiffeners are extruded 6061-T6, while the plate making up the container consists of 5454-H32. Again, the vast majority of the joints are welded; with only the corner fittings joined using mechanical fasteners. As the case with rail operations, weight savings over a fleet of similar steel containers results in economic benefits. Additional product may be transported each time a container is moved, thereby reducing the number of trips or containers required. A smaller number of containers translate into reduced labor costs.

Figure 7 – *Intermodal Shipping Container (Courtesy of Aerospace and Commercial Rolled Products Group, Alcoa)*

Outlook and Future Research Needs

Research needs, as defined by the aluminum industry, have recently been summarized (ENERGETICS, 1997). Broken down into the primary product sector, casting, rolling and extrusion, and finished products, not all are directly related to material needs of the infrastructure. An attempt to highlight those having the largest impact on the infrastructure follows.

From the primary product sector, one of the major goals is the reduction of energy required to produce aluminum in the smelting process. While aluminum is an abundant metal, extraction and reduction is expensive. Disposal of byproducts, like salt cakes and spent pot-linings, tend to elevate finished product costs. New processes which eliminate wastes, or new applications for byproducts would assist with overall cost reduction. Lowering costs of primary aluminum as well as finished mill products would make the material more attractive to owners and operators. Despite efforts to examine life cycle costs, no universally accepted methodology exists. Hence, first cost is still utilized when awarding contracts. A related issue is the treatment of recycled materials. Currently, only like alloy types may be re-melted and cast together. Improvements to sorting, separation and cleaning would assist with the minimization of impurities. Increased utilization of scrap material would lower costs and enhance competitiveness.

While near net shape casting processes have made significant improvements over the last several years, potential for even greater impact would be development of continuous cast sheet and plate for the entire family of aluminum alloys. The ability to continuously cast sheet and plate would eliminate processing steps required to arrive at a finished rolled product. Elimination of process steps would assist with the overall goal of reducing the costs of finished goods, making use of aluminum in the infrastructure economically more attractive.

Barriers to the development of alloys with improved properties include the lack of fundamental, detailed understanding of the process - microstructure - property relationship. Alloy enhancement and process optimization requires development of improved computational models and experimental techniques. For example, models and physical experimentation could be employed to develop a better understanding of metal flow in an extrusion die, improve die design and produce a wider variety of extruded alloys at increased production rates. In addition, fundamental understanding of the rolling and reduction processes, as well as the influence of thermal history could be used to develop enhanced formability characteristics in sheet materials.

Among the goals for end products is the need to extend the useful life of aluminum components and structures. One possible avenue is through new and improved alloys. Enhanced corrosion resistance as well as alloys possessing higher elastic modulus values would be of significant benefit. Enhanced corrosion resistance would certainly extend the life of components and structures in severe environments. An improved elastic modulus would impact both functional and serviceability requirements. An increase in modulus among geometrically similar components would enhance stiffness, increase natural frequencies, lower overall deformation and enhance crack growth resistance.

Design specifications should be developed to cover additional aluminum materials. One prime example is castings. The majority of the work used to develop the current Aluminum Association Design Manual provisions were conducted on wrought products (Aluminum Association 1994). As such, the provisions are not necessarily adequate for cast components. Another area requiring development is plastic design, allowing full use of a member's cross section to resist loading. Static welded joint design may be too conservative as historic test results indicate significantly larger resistance than allowed by the specification. In other instances, not enough procedural guidance is provided. An example would be proportioning of mechanically fastened joints and consideration of applicable limit states. Two such areas requiring research include shear lag and tear out failures of bolt groups.

Consistent with extending the life of aluminum products is improved understanding of fatigue resistance of materials, joints and members. Certainly, increasing the modulus would be expected to enhance crack growth resistance. One primary shortcoming of current specifications is the lack of design data on a wide variety of full-scale joints. Versatility of aluminum products often leads to new or unique details that are not easily classified. The knowledge base should be expanded to a wider variety of joints. Another possibility would be to employ a methodology that eliminates classification and promotes use of advanced analytical design techniques. The hot-spot technique is but one possible alternative. Extension of product life also implies design in the high cycle regime. Additional data and variable amplitude studies are required.

Conclusions

Future prospects for utilization of aluminum in the infrastructure will increase, given the physical characteristics, flexibility of product forms and relative ease of fabrication. Efforts to reduce first costs should continue, and result in more competitive initial cost. Aluminum alternatives should be examined more favorably as life cycle cost methodologies become accepted and mature.

Material development has been accepted as a market specific process for some time, and is likely to continue in this manner. While alloys have been developed for specific markets, wider application may be found. An example would be the marine based 5XXX series materials utilized for building of surface effect ships, transport storage tanks and high speed lightweight vessels. Many of these materials possess relatively high strengths, exceptional corrosion resistance and are readily weldable. Many would find application to the rebuilding of the infrastructure. The ability to utilize different alloys and product forms results in maximum flexibility for the designer. Material may be placed where needed, resulting in unique cross sections and structural form.

Material research should concentrate on cost-effective ways to develop improved elastic properties while maintaining features attractive for the designer. Some important items would include enhanced corrosion resistance, increased elastic modulus, improvements to abrasion resistance and changes to elevate annealed strengths. Coupled with material research are items critical to further advancement of design provisions. Uniform rules for handling cast components, improved methods for plastic design, mechanical joint proportioning, realistic evaluation of the static strength of welded joints, improved evaluation techniques for tapered segments of cross sections and complex shapes and improved fatigue design rules are needed.

Acknowledgments

The compilation of this work would not have been possible without the support of The Department of Civil Engineering at The University of Akron. Special thanks are due to David Bennett, Dan Roup and Brian Foy of the Aerospace and Commercial Rolled Products Division of Alcoa, Jim Burg of the Alcoa Technical Center, and Gil Kaufman of The Aluminum Association.

References

Aluminum Association, (1980),"Aluminum for More Efficient Railroads", First Edition, The Aluminum Assoication, Washington, DC.

Aluminum Association, (1988), "Aluminum Standards and Data", Ninth Edition, The Aluminum Association, Washington, DC.

Aluminum Association, (1991), “Welding Aluminum: Theory and Practice”, Second Edition, The Aluminum Association, Washington, DC.

Aluminum Association, (1994), “Aluminum Design Manual”, First Edition, The Aluminum Association, Washington, DC.

Aluminum Association, (1995), “Automotive Aluminum Recycling”, Brochure, The Aluminum Association, Washington, DC.

Aluminum Association, (1996), “Aluminum Alloys for Bridges: Extending the Life of U.S. Bridges”, The Aluminum Association, Washington, DC.

Aluminum Association, (1997), “International Alloy Designations and Chemical Composition Limits for Wrought Aluminum and Wrought Aluminum Alloys”, The Aluminum Association, Washington, DC.

Bergsma, S. C., Kassner, M. E., Li, X., Delos-Reyes, M. A. and Hayes, T. A., February (1996), “The Optimized Mechanical Properties of the New Aluminum Alloy AA 6069”, Journal of Materials Engineering and Performance, American Society for Metals, Vol. 5.

Boylan, J.,(1997), ”Semi-Solid Forged Aluminum Components for High Stress Structural Automotive Applications”, Proceedings of Alumitech ’97, The Aluminum Association, Atlanta, Georgia.

DURALCAN, (1996), ”Composites for Wrought Products – Property Data”, Product Brochure, Novi, Michigan.

DWA, (1997), “Company Profile – DRA Product Catalogue”, DWA Aluminum Composites, Chatsworth, California.

ENERGETICS, (1997), “Aluminum Technology Roadmap Workshop”, Sponsored by the Office of Industrial Technologies U.S. Department of Energy, ENERGETICS, Inc., Columbia, Maryland.

ESAB, (1997), “ESAB Super Stir - Friction Stir Welding Machines and Installations for Weldjoint in Aluminum”, Product Brochure, ESAB Welding Division AB, Laxa, Sweeden.

EWI, (1996), “Friction Stir Welding at EWI”, Friction Stir Welding Flyer, Edison Welding Institute, Columbus, Ohio.

Hart-Smith, L. J., (1987), “Design of Adhesively Bonded Joints”, Joining Fiber Reinforced Plastics, Edited by F. L. Matthews, Elsevier, New York.

Hatch, J. E., Editor, (1984), "Aluminum: Properties and Physical Metallurgy, American Society of Metals, Metals Park, Ohio.

Hinkle, A. J. ,Brockenbrough, J. R. and Burg, J. T., (1996),"Microstructural Material Models for Fatigue Design of Castings", Society of Automotive Engineers, SAE #960161, International Congress and Exposition, Detroit, Michigan.

Hval, M., (1997), "A New High Strength Aluminum Alloy for Marine Application", Proceedings of Alumitech '97, Sponsored by The Aluminum Association, Atlanta, Georgia.

Kosteas, D. and Sanders W., (1995), "Redesigning an Aluminum – Concrete Composite Bridge", Proceedings of the Sixth International Conference on Aluminum Weldments", American Welding Society, Cleveland, Ohio.

Kvale, I., Vallestad, I. J. and Solaas, K., (1997), "Forsmo Bridge – The First Aluminum Road Bridge in Norway", Proceedings of Alumitech '97, Sponsored by The Aluminum Association, Atlanta, Georgia.

Menzemer, C. C., Hinkle, A. J. and Nordmark, G. E., (1987), "Aluminum Orthotropic Bridge Deck Verification", Materials and Member Behavior, Structrures '87 Proceedings, ASCE, Orlando, Florida.

Progressive Railroading, February (1986), "Electric Companies Go For Aluminum", Progressive Railroading.

Raynaud, G. M. and Gomiero, P., (1997),"The Potential of 5383 Alloy in Marine Applications", Proceedings of Alumitech '97, Sponsored by The Aluminum Association, Atlanta, Georgia.

Sampath, S., Schipper, H., Haszler, A. and Weber, G.,(1997), "Development of Advanced Ship Building Materials at Koninklijke Hoogovens N. V.", Proceedings of Alumitech'97, Sponsored by The Aluminum Association, Atlanta, Georgia.

Sharp, M.L., Nordmark, G. E. and Menzemer, C. C., (1996), "Fatigue Design of Aluminum Components and Structures", McGraw-Hill.

Zehe, R., (1997), "The Maturation of Semi-Solid Processing", Proceedings of Alumitech '97, Sponsored by The Aluminum Association", Atlanta, Georgia.

Karim S. Rebeiz[1]

Chapter 5

Polymer Concrete Made with Recycled Plastics

Abstract

The utilization of high performance materials in engineering applications has emerged as an issue of considerable importance. Indeed, there has been a distinguished tradition among academics to disseminate information about high performance materials in the prevailing literature. Still, the use of high performance materials remains to be fully established and accepted in industry. The main obstacle to the widespread utilization of high performance materials is their high cost relative to conventional materials. This chapter describes the use of polymer concrete (PC) made with recycled materials. In this context, the primary use of recycled materials is not to alleviate an environmental problem posed by waste materials, although this is a very noble objective. Rather, their utilization is warranted to achieve an optimum cost of PC without compromising the integrity or performance of the material.

Keywords

Polymer concrete; recycled plastics; fly ash; PET; properties; steel-reinforcement.

Introduction

The development of new high performance materials that are stronger, more durable and more efficient to apply than conventional materials is an important consideration in many engineering applications. Polymer concrete (PC) is an example of such a relatively new technology that has very promising uses in construction, civil and other engineering applications (Guide for the Use of Polymers in Concrete 1997; Ohama et al. 1997). PC consists of well-graded inorganic aggregates bonded together by a polymer binder. PC is produced by using a resin binder for the inorganic aggregates rather than the conventional water and cement binder used in portland cement concrete. The most common binder used in PC is unsaturated polyester because of its good properties and relatively low cost compared to other resin binders (Rebeiz 1991). Examples of other types of resins used in PC production include methyl methacrylate, polyesteramide , epoxy and vinyl ester.

PC is much stronger and more durable than portland cement concrete (Fowler 1999; Chandra and Ohama 1994). Most importantly, it cures much faster than

---

[1] ASCE Member, Ford Motor Company, Dearborn, MI.

cement-based materials. Typically, it takes a few minutes or hours for PC to cure at ambient temperature, whereas portland cement concrete cures in days or weeks. However, the use of PC is still limited because of the lack of familiarity of the product by the practitioners, and its perceived high cost relative to conventional materials. The intent of this chapter is twofold: To introduce PC materials as a feasible and viable emerging material in engineering applications; and to demonstrate that the production of cost effective, high quality PC materials could be achieved using recycled products without compromising the performance and integrity of the material. Developing high performance PC products using recycled materials is of particular importance to various civil engineering, construction, manufacturing, governmental and environmental entities. In this context, a high performance material is defined as having the ability to achieve special performance requirements under routine formulation and placement practices that cannot otherwise be obtained via the use of conventional materials. PC, like other high performance materials, is still in its infancy in the sense that it still does not enjoy an established and formalized sets of codes and standards that regulate the formulation, design, testing and placement of the material.

The two types of recycled materials that have been used in the production of PC are recycled poly(ethylene terephthalate), PET, mainly recovered from used plastic beverage bottles, and fly ash, a by-product of burning coal to produce energy. The infrastructure to collect used plastic bottles has been implemented successfully in many localities. Thus, a relatively large and easily accessible waste stream of recycled PET is available for recycling applications (Rebeiz and Craft 1995a). The recycled PET is chemically modified to produce unsaturated polyester resin; in turn the polyester resin is mixed with aggregates to produce PC. The fly ash waste is used as an inert filler to improve the mechanical properties and the surface appearance of the PC (Rebeiz et al. 1999; Rebeiz et al. 1996a; Rebeiz et al. 1996b; Rebeiz et al. 1995b). The disposal of fly ash can pause some environmental problems because the material may often be considered hazardous. PC is an excellent medium to contain potential leaching of fly ash (if such a leaching might be considered hazardous) because of the inherently very low permeability and excellent chemical resistance of the PC materials.

Emerging Material

Polymers

There is no doubt that the use of polymers is expected to grow dramatically in industries as diverse as the construction, electrical and automotive industries (Fowler 1999; Rebeiz et al. 1993a; Omaha et al. 1997; Gunasekaran 1997). The main advantage of polymer-based products is their versatility in terms of formulations, properties and performances. Indeed, there are more than 50 families of polymeric materials, each possessing unique characteristics (Mark 1990). For example, two polymers can be as different as steel and wood. The major families of polymers are polyethylene, polyvinyl chloride, polystyrene, polyester, and polypropylene.

A polymer is made of macromolecules or long molecular chains consisting of small monomers (single units) joined together by covalent bonds. Polymers can be in the form of viscous liquids, soft elastic rubbers, or hard solids depending on their molecular structure and the ambient temperature. Most polymers are organic, that is they are based on a carbon chain. Inorganic polymers are based on a backbone chain other than carbon, mainly silicon. Polymeric materials may occur naturally or be manufactured synthetically. Examples of natural polymers are wood, leather, silk, and asphalt. Synthetic polymers include rubbers, coatings, plastics, and adhesives.

Three basic structures may occur in a polymer: linear, branched, and cross-linked network. A linear polymer is one in which the skeletal structure consists of a long chain of atoms to which the side groups are attached. A branched polymer has side branches of similar structure than the backbone chain. A cross-linked polymer is one where very strong bonds (cross-links) exist between the polymer chains to form a three-dimensional network (Mark 1990).

Polymers can be grouped into two broad categories: Thermoplastics and thermosets. Thermoplastics are polymers that can soften and flow when heated without damaging their properties. Conversely, thermosets are polymers that cannot soften and flow when heated without causing major damages to their properties. Thermosets cannot be reprocessed because very strong bonds exist between the molecular chains (mainly in the form of chemical cross-links) that will prevent them from moving apart from each others due to the action of heat or solvents.

Three methods are used to produce polymer based-products. The first method consists of heating a thermoplastic to temperatures at which it can flow quite readily. The polymer is then extruded, injected or sprayed depending on the final use of the product. The thermoplastic then hardens as it is cooled down. Examples of products that can be produced using this method are pipes, buckets, or joints. The second method consists of applying a resin, which has been diluted in a solvent. After application, the solvent evaporates, allowing the resin to cure. Adhesives and coatings can be grouped into this category. The third method consists of applying monomers or resins that are in a partially polymerized state. They are mixed with activators (initiators and promoters for example) and then applied. The polymerization is completed in-situ. PC is an example of such a system that is being described subsequently.

Polymer Concrete

It is important to realize that PC is not a single material with fairly defined formulations and characteristics. Instead, it is a very versatile material with a wide range of properties and performances. This versatility does not only apply across resins; it also applies to resins within the same family. For example, unsaturated polyesters do not represent a single family of materials with similar characteristics; instead, unsaturated polyester resins are a large family of materials, each with its own characteristics and performances. For example, an unsaturated polyester resin could be formulated to have low viscosity, low modulus and high elongation at break for PC overlay applications on portland cement concrete with large temperature variations; conversely, another unsaturated polyester can be formulated for high strength and high modulus for structural precast applications (Rebeiz et al. 1995c; Banko et al. 1994; Rebeiz et al. 1992).

PC may be used effectively as thin overlays on portland cement concrete pavements and bridges (Fowler 1999; Doody and Morgan 1994). The overlay protects the portland cement concrete substrate by providing low water and chloride permeability. It also acts as a surfacing material to repair spalled and deteriorated portland cement concrete and maintains high skid resistance, thus improving ride quality and safety. The material may also be used as a precast product in many utility, transportation, and electrical and diversified manufacturing-related components (Fowler 1999; Boneti et al. 1999; Ohama et al. 1997; Bloomfield 1997; Karcas 1997; Gunasekaran 1997; Chandra and Ohama 1994). Examples include vaults, junction boxes, trench lines, floor drains, sewers and manholes, bridge panels, railroad ties, high voltage insulators, containers for chemical wastes and machine bases. Other uses of PC include durable, attractive floor blocks or tiles, especially for industrial applications. Their high strength, impact resistance, chemical resistance, and low permeability make them resistant to heavy traffic, chemicals, and water. They also act as a shock absorber and reduce the noise and vibration of heavy equipment. Building panels could also be produced as both single skin and sandwich panels. The ability to easily add color, to form various shapes, and to use exposed aggregate finishes make PC a very feasible material for architectural precast concrete. PC could also find applications as adhesives for joining different components of same or different composition. For example, PC adhesives could be used to bond new concrete to old concrete, or to bond old concrete to steel plates, steel bars, or heavy equipment machinery (Guide for the Use of Polymers in Concrete 1997). Low modulus, high elongation resins should be used in joints where movement must be allowed.

Research Needs

The most pressing research need for PC is the formulation of resins that achieve optimum performance at the lowest possible cost. Indeed, most of the PC cost comes from the resin component; the cost of the filler component is comparatively negligible. If resins can be produced at a lower cost, then their use may significantly enhance the potential use of PC-based products. During the last decade, some work has been done on the production of unsaturated polyester resins based on recycled PET. (Vaidya and Nadkarni 1987). The polyester resins could then in turn be mixed with inorganic aggregates for PC production (Rebeiz et al. 1991 and 1990). Recycled PET is a plastic waste mainly recovered from used beverage bottles. A large waste stream of PET is available for recycling applications as many localities are recovering beverage plastic bottles after use. The recycled PET could also be used in a number of other products such as coats, pillows, or carpets. However, a main advantage of recycling PET in PC is that the PET materials do not have to be purified, including removal of colors, to the extent other PET recycling applications might require, which should facilitate the PET recycling process and making it more cost efficient. The production of unsaturated polyester resins based on recycled PET is discussed subsequently.

Formulation of Polymer Concrete Using Recycled PET and Fly Ash Waste

The Chemical Modification of Recycled PET for Unsaturated Polyester Production

PET consists of repeating ethylene glycol (EG) and terephthalic acid (TPA) connected together through ester linkages. For the production of unsaturated polyester, the PET molecules are converted into low molecular weight oligomers by glycolysis in the presence of a transesterification catalyst (Rebeiz et al. 1991; Vaidya and Nadkarni 1987). The reduction of PET into short chain segments may be done by charging PET, propylene glycol (PG), and catalyst into a glass reactor and heating under a nitrogen atmosphere. As shown in Figure 1, this is a reversible reaction where the free PG replaces EG from the polymer chain by a process of chain scission and glycol exchange. A dibasic acid, typically maleic anhydride (MA), is then be added to the solution to produce unsaturated polyester. The unsaturated polyester condensation from glycolysis products and MA is shown in Figure 2. The water by-product is removed throughout the course of the reaction. The unsaturated polyester resins that have been formed now are prepolymers with high viscosity. They are therefore diluted with a solvent monomer to impart certain specific characteristics. Styrene is an example of such a solvent. Styrene reduces resins viscosity to make them workable and allows their further cure to a solid (polymer) by participating in the crosslinking via chain-reaction (addition) polymerization upon addition of free-radical initiators and promoters. As much as 50% styrene may be used in unsaturated polyesters without adversely affecting the physical and mechanical properties of the hardened materials (Ohama et al. 1979).

Different formulations and different percentages of recycled PET may be used in making the polyester resins. The percentages of recycled PET used in these resins could vary from a low concentration of 15% to a high concentration of 40% (Rebeiz et al. 1990). The maximum possible percentage of recycled PET is usually desirable in making unsaturated polyesters because the PET residues do not adversely affect PC properties and their use helps in reducing the cost of the material (Rebeiz 1994). A main advantage of using recycled PET in making unsaturated polyesters, as compared to using virgin materials, is that it takes about 50% shorter processing time to produce a polyester resin with a certain molecular weight and acid number (Rebeiz et al. 1991).

PC made with virgin resins typically utilize isophthalic or orthophthalic-based polyesters. Conversely, resins based on recycled PET are terephthalic-based polyesters. Terephthalic-based polyester resins could produce a more uniformed and structured cross-linking matrix with a higher degree of strength, stiffness, and toughness than isophthalic or orthophthalic-based polyesters because of the location of the carboxyl groups on the benzene ring of phthalic acid. It is worth mentioning that virgin terephthalic acid is not usually used in the production of unsaturated polyester because it is expensive and possesses a high melting point, which presents difficulties in synthesis. That is the reason why recycled PET is an effective method of incorporating terephthalic functionality into the backbone of a polyester resin.

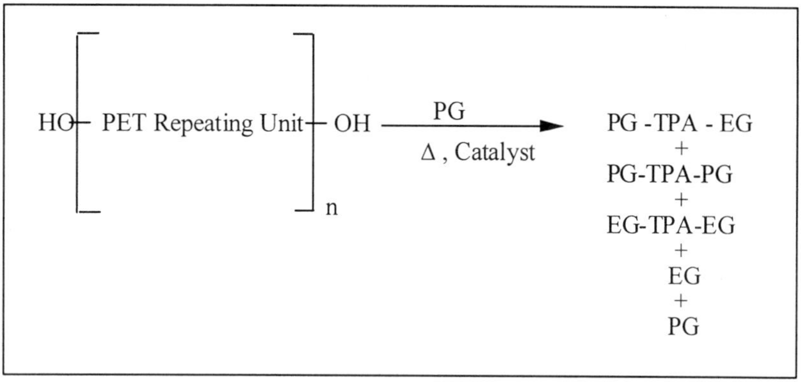

Figure 1 - *The Glycolysis of the Recycled PET*

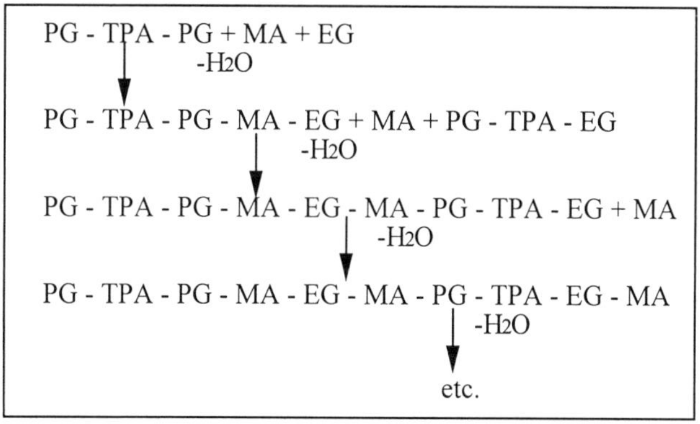

Figure 2 - *The Condensation of the Unsaturated Polyester*

Polymer Concrete Production

The production of PC using unsaturated polyester resins based on recycled PET is illustrated in Figure 3. Initiators and promoters are added to a liquid polyester resin to initiate the curing process. A typical initiator is methyl ethyl ketone peroxide and a typical promoter is cobalt naphthanate. Immediately after the addition of initiator and promoter chemicals, the resin is mixed with gravel, sand and fly ash and allowed to cure at room temperature. Inorganic aggregates should be oven-dried to ensure good adhesion between the polymer matrix and the aggregates. Curing takes place because styrene combines with the reactive double bonds of the polyester chains, thus linking them together and forming a strong three-dimensional polymer network.

The typical physical properties of two unsaturated polyester resins based on recycled PET, especially formulated for PC use, are shown in Table 1. Resins 1 and 2 used 40% and 30% by weight, respectively, of recycled PET in the alkyd portion of their formulations, i.e., before the addition of styrene. Resin 1 was dark green in color whereas resin 2 was clear amber since the recycled PET used in the production of resin 1 was not purified to the same extent as the one used in the production of resin 2. In most PC applications, the purity of the recycled PET is not very important, which should result in a decreased cost for unsaturated polyester resins based on recycled PET. Resin 1 had low viscosity and excellent wetting properties for the inorganic aggregates; it is therefore suitable for the production of PC using a high aggregate-to-resin ratio. A high aggregate-to-resin ratio is not only desirable for economy (since less resin, the expensive component, is used), but it also improves the dimensional stability due to load, temperature or shrinkage, the chemical stability and resistance, and the flame spread resistance of the PC.

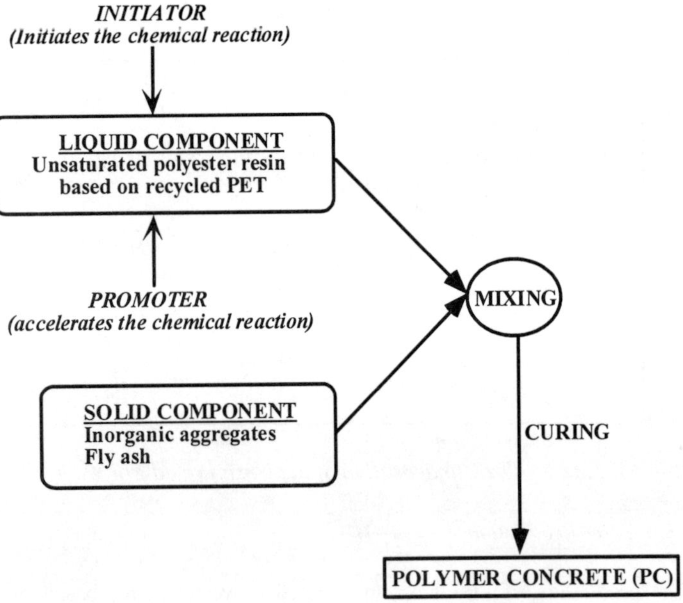

Figure 3 - *The Production of PC*

Table 1 - *Physical Properties of Two Resins Based on Recycled PET*

Properties	**Resin 1**	**Resin 2**
Weight percent of recycled PET in alkyd portion (%)	40	30
Weight percent of styrene monomer in resin (%)	40	30
Viscosity at 25 °C (cps)	110	700
Weight average molecular weight	3340	8000
Number average molecular weight	1000	1500
Acid number	8	20
Specific gravity	1.09	1.16
Appearance	Dark green	Clear amber
Tensile strength (MPa)	21.1	13.3
Tensile modulus (GPa)	1.24	0.30
Elongation at break (%)	2.4	53.5

The mix design for the PC systems should be optimized for workability, strength, and economy depending on the intended application. For example, a mix especially formulated for precast applications consists of an aggregate-to-resin ratio of 9-to-1: 10% resin, 45% oven-dried pea gravel, 30% oven-dried sand, and 15% fly ash. Conversely, another mix especially formulated for thin overlay applications uses an aggregate-to-resin ratio of 4-to-1: 20% resin, 60% oven-dried sand, and 20% Class F fly ash. The use of fly ash greatly improves the workability of the fresh mix. The fine and spherical particles of fly ash provide the fresh mix with better lubricating properties, thus improving its plasticity and cohesiveness. The better gradation obtained with fly ash also results in a hardened material with improved strength properties and surface appearance.

Properties of PC using Recycled PET and Fly Ash Waste

Strength Properties

The neat resins (cured resins without the use of aggregates) are first examined for their tensile properties since such an evaluation has great influence on the potential use of the PC material. The typical tensile stress-strain curves for two neat resins based on recycled PET are shown in Figure 4. It can be observed in the figure that the stress-strain properties of the two neat resins are quite different. One resin is rigid since it has a high tensile strength, high tensile modulus, and low elongation at failure. Conversely, the other resin is flexible as evidenced by the fact that it has a low tensile strength, low modulus, and high elongation at failure. These different properties are very important in the potential use of these resins in PC applications. Rigid resins are suitable in precast applications because they produce high modulus PC structures that deflect and creep less under the action of loads; conversely, flexible resins are suitable in PC overlay applications because they are capable of stretching significantly before breaking when subjected to large thermal or mechanical movements (Rebeiz et al. 1993b).

Typical stress-strain curves in compression for two PC systems, one using a flexible resin and the other using a rigid resin, are shown in Figure 5. It can be seen that the type of resin (rigid or flexible) used in making PC can significantly affect the stress-strain properties of the material. PC made with a flexible resin exhibits lower modulus, higher ductility, and more toughness than the PC made with a rigid resin. The stress-strain behavior and the ultimate compressive strength of PC using resin based on recycled PET are comparable to those typically obtained with PC using virgin resins (Vipulanandan 1987). It can also be observed that the ultimate compressive strain of PC is much larger than the one corresponding to portland cement concrete.

The properties of the PC are strongly influenced by the type, properties, and amount of fillers. The amount of fillers could vary from 10% to 90% by weight depending on the resin and filler properties, the intended use of the polymer composite, and how it is to be processed. An insufficient amount of fillers disrupts the material homogeneity and hence decreases the strength of the PC. However, an increase in the amount of filler beyond a certain level increases the strength of the PC to a value exceeding the strength of the neat resin (high-filler effect). The maximum amount of filler is approximately 90%. Beyond this maximum value, the resin content is not enough to coat the fillers and the mechanical and physical properties of the PC are sharply lowered.

Figure 6 shows a plot of the typical compressive stress versus the axial and lateral strains for two PC systems using different aggregate-to-resin ratios. The PC system with a relatively high aggregate-to-resin ratio exhibits significantly higher compressive modulus and lower Poisson's ratio than the PC system with a relatively low aggregate-to-resin ratio because inorganic aggregates have a higher modulus and lower Poisson's ratio than the resin binder. Accordingly, the level of aggregate-to-resin ratio in the mix design, as well as the type of resin itself significantly affect the compressive modulus and Poisson’s ratio of PC.

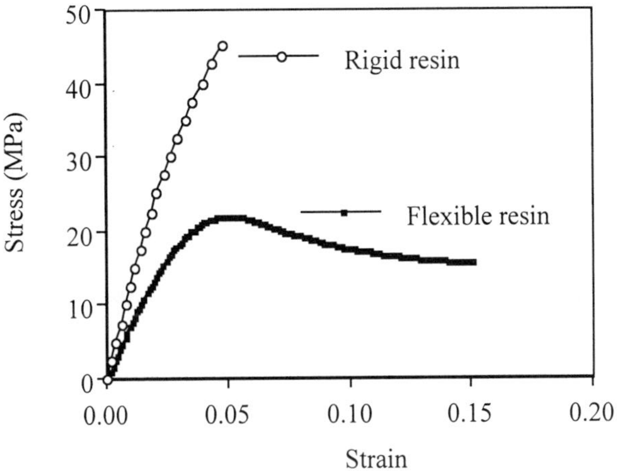

Figure 4 - *The Typical Tensile Stress-Strain Curves for the Neat Resins*

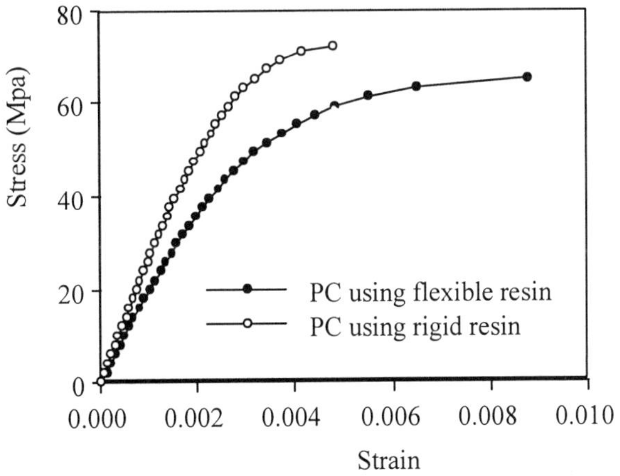

Figure 5 - *The Typical Compressive Stress-Strain Curves in Compression for PC*

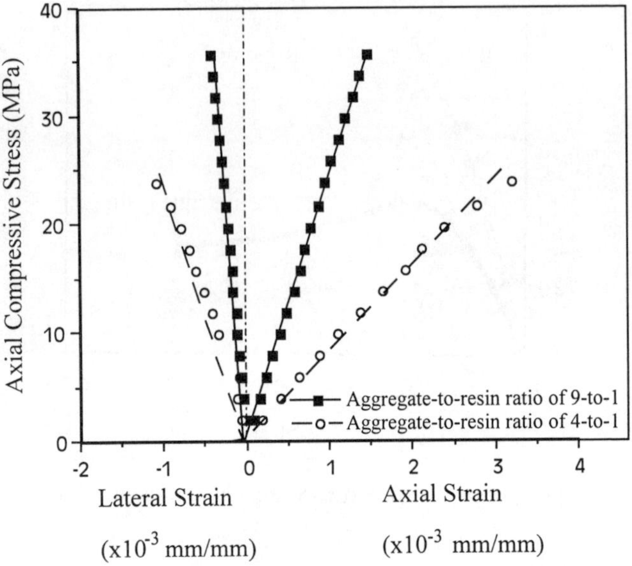

Figure 6 - *The Typical Axial-Lateral Strain Under Compressive Stress for PC*

Curing Characteristics

The curing characteristics (shrinkage and exotherm) are important in many PC applications. In precast components, low shrinkage is important because excessive shrinkage strains may significantly affect the dimension of these structures, thus making their demolding, assembly, or use more difficult. In overlay applications, relatively low shrinkage is desirable because studies have reported that excessive shrinkage strains may cause delamination between the overlay and the substrate (Al-Negheimish 1988). Figure 7 shows the curing characteristics (shrinkage and exotherm curves) for two PC systems. The shrinkage strains are measured during the polymerization of the PC materials using a special device (Rebeiz et al. 1992). As shown in Figure 7, the ultimate shrinkage strain in the PC system using a weight aggregate-to-resin ratio of 4-to-1 is about three times higher than the one corresponding to the PC system using a weight aggregate-to-resin ratio of 9-to-1. The reason for this occurrence is because a larger amount of resin is susceptible to shrinkage in a low aggregate-to-resin ratio than in a high aggregate-to-resin ratio. Most of the shrinkage in both PC systems takes place within the first eight hours after mixing and stops beyond 24 hours.

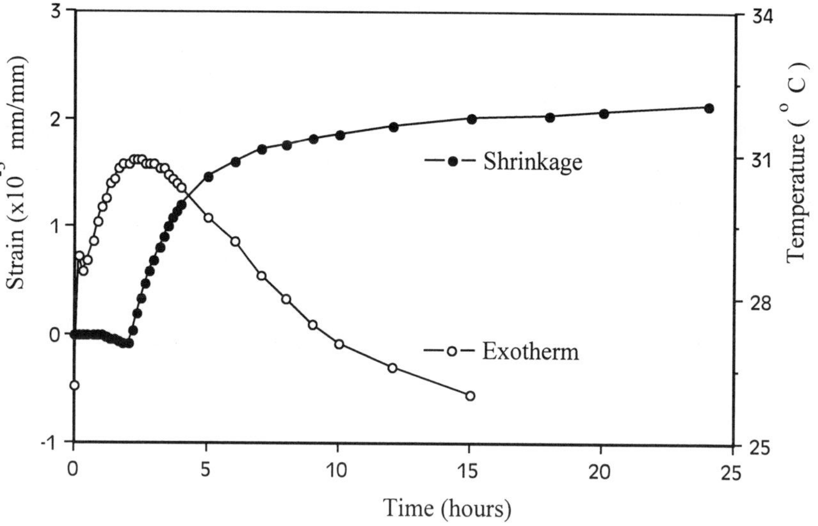

(a) Aggregate-to-resin ratio of 9-to-1

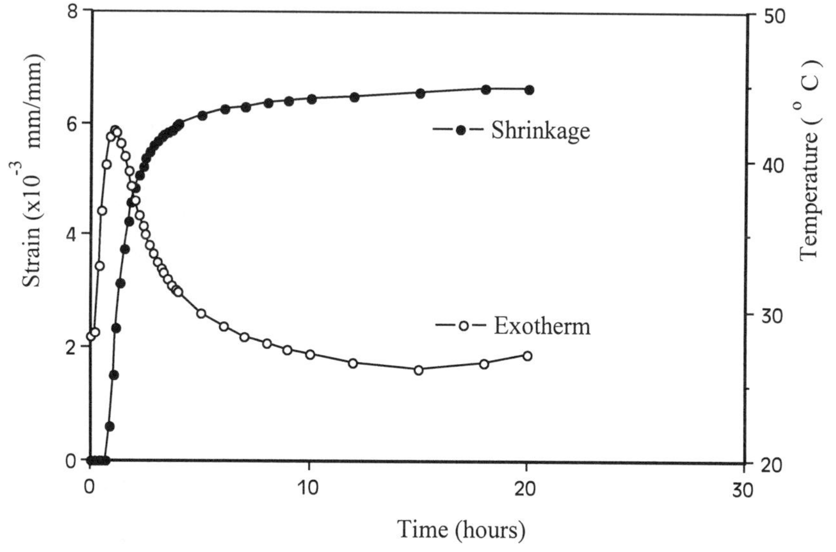

(b) Aggregate-to-resin ratio of 4-to-1

Figure 7 - *The Typical Shrinkage and Exotherm Curves for PC*

The peak exotherm occurs approximately two hours after casting. The time of occurrence of the peak exotherm is independent of the aggregate-to-resin ratio. However, the magnitude of the peak exotherm for an aggregate-to-resin ratio of 4-to-1 is significantly higher than the one corresponding to an aggregate-to-resin ratio of 9-to-1. This finding is not surprising since a higher level of polymerization takes place in a low aggregate-to-resin ratio than a high aggregate-to-resin ratio. The exothermic reaction for the PC systems ends completely 10 hours after casting as evidenced by the flattening of the curves.

Thermal Expansion

Thermal expansion is important when PC materials are used in conjunction with other materials such as steel or portland cement concrete since the coefficient of thermal expansions of PC are significantly higher than those corresponding to steel or cement concrete. Hence, changes in temperature in the composite structure will create relatively high shear stresses at the interface between the two materials, which may eventually cause deterioration in the structure. Figure 8 shows the thermal expansion for two PC systems using different aggregate-to-resin ratios: One using a weight aggregate-to-resin ratio of 9-to-1 whereas the other using a weight aggregate-to-resin ratio of 4-to-1.

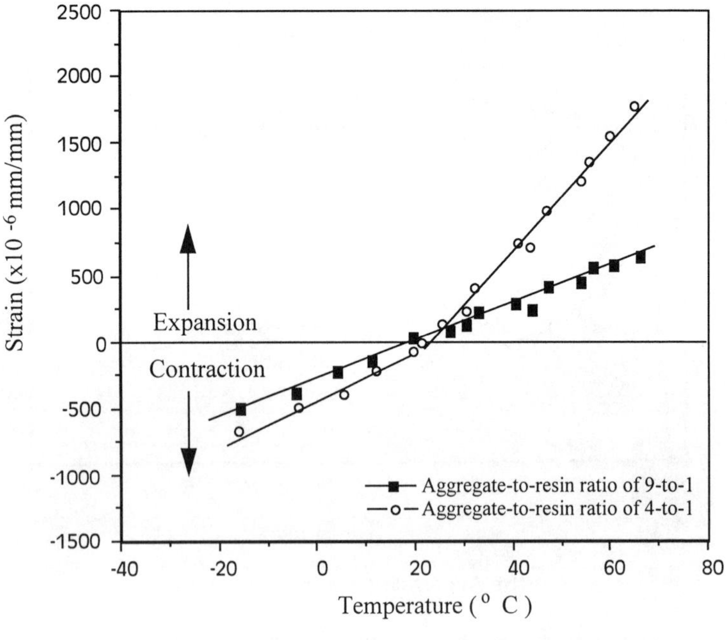

Figure 8 - *The Typical Thermal Expansion Curves for PC*

As shown in Figure 8, the coefficient of thermal expansions of PC are at least twice as high as those corresponding to steel or cement concrete. The amount of filler in the PC significantly affects the thermal behavior of the material. PC using a weight aggregate-to-resin ratio of 9-to-1 has a lower thermal expansion than the PC using a weight aggregate-to-resin ratio of 4-to-1 because inorganic aggregates are less sensitive to variations in temperature than the resin binder. In addition, the PC using a weight aggregate-to-resin ratio of 4-to-1 exhibits a sudden change in thermal expansion characteristics as the temperature exceeds 20 °C, which is not the case with the PC using a weight aggregate-to-resin ratio of 9-to-1. Another study shows similar behavior with PC using virgin materials (Al-Negheimish 1988).

Creep

A common problem encountered with PC systems used in precast components is excessive creep deformation under service conditions. Creep takes place in PC as the result of molecular movement in the viscoelastic resin binder. Because of the viscoelastic resin binder, variations in time and temperature significantly affect the PC creep behavior. A typical creep compliance (or strain per uniaxial unit sustained stress) and Poisson's ratio versus time for PC are shown in Figure 9.

A stress intensity ratio of 20% was chosen in the creep evaluation test to avoid complications resulting from nonlinear viscoelastic behavior and because PC is usually designed with a high safety factor.

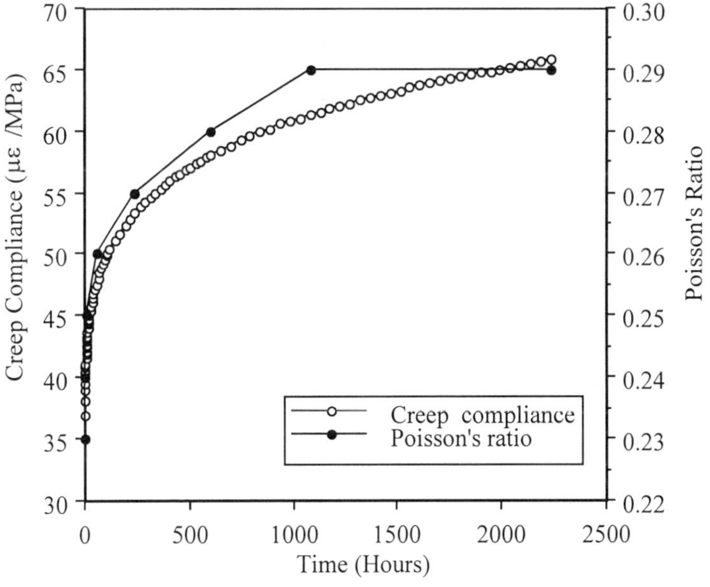

Figure 9 - *The Typical Creep Compliance and Poisson's Ratio for PC*

The application of larger load levels would also have been more difficult because of the high strength of PC (Kyriacou 1991). More than 20% of the final creep for PC takes place within one day and more than 90% of the final creep occurs in six days.

The specific creep (which is the creep strain divided by the sustained stress) is 28 me/MPa after 95 days, which is comparable to what was observed with other PC systems using virgin resins (Hristova and Bares 1987). It has also been established that the changes in the creep compliance of the PC adhere to the principle of the time-aging time equivalence with the reduction function depending on aging temperature (Aniskevich and Hristova 1999). Previous creep studies with PC determined that excessive creep deformation and catastrophic failure often occur when the creep stress intensity ratio (which is the ratio of the applied compressive stress to the ultimate compressive strength) exceeds about 50 percent (Hsu and Fowler 1985). Compared to cement concrete, PC ultimate creep strain is about one to two times higher, while the creep compliances of both materials are about the same. It should be noted, however, that the creep behavior of PC is sensitive to temperature variations while the creep behavior of cement-based materials is sensitive to humidity changes.

There are different factors that affect the creep behavior of PC. The most important one is the molecular structure of the resin. An increase in the cross-linking or molecular weight in the resin results in a reduction of creep deformations by restricting the amount of slip between the molecules (Moore and Kline 1984;. Alkonis et al. 1972). This fact explains why thermoplastics tend to generally undergo larger creep deformations than thermosets. Another factor that affects PC creep behavior is the chemical nature of the resin. For example, PC systems using polyester resins tend to creep more than PC systems using epoxy resins (Ayyar and Deshpande 1981). A third factor affecting PC creep behavior is the mix design. PC systems made with a high aggregate-to-resin ratio exhibit lower creep deformations than PC systems made with a low aggregate-to-resin ratio (Ayyar and Deshpande 1981; Dharmarajan and Armeniades 1987). Inorganic aggregates have a relatively high stiffness and therefore tend to deform less than the resin binder. However, as the aggregate-to-resin weight ratio exceeds 5-to-1, the effect of the resin binder on the creep behavior becomes minimal (Hristova and Bares 1987). The use of very fine aggregates, the addition of reinforcements, and the use of shrinkage reducing agents have been found to increase the creep resistance of PC (Moore and Kline 1984; Hristova and Bares 1987; Nutt, W. O., and Staynes 1987; Armeniades, C. D., Dharmarjan 1987). For example, it was found that calcium carbonate, a micro-filler, can reduce the creep deformation of PC by about 10 to 15 percent (Mani 1983). However, the addition of coupling agents used to increase the bond between inorganic aggregates and the resin binder have no apparent effect on the creep behavior of polyester composites (Mani 1983). A final factor affecting PC creep behavior is the extent of curing. As the degree of polymerization in the resin increases, the creep deformation in the PC decreases. After polymerization is completed, the age at loading for PC does not have an effect on its creep behavior (Dharmarajan and Armeniades 1987).

Durability Properties

The durability properties of PC are shown in Table 2 (the values in the table represent the average of 5 specimens). Good chemical resistance is important when PC overlays are used to protect industrial floors, tanks, pipes or containers from aggressive liquids or radiations from nuclear wastes. In very aggressive environments, especially formulated polyester resins is used in making PC to provide the necessary resistance against the action of heat and flame, strong chemicals, or other aggressive agents. It should be noted that unsaturated polyesters have usually relatively good resistance to acids, while they have relatively poor resistance to alkalies.

As shown in Table 2, PC has low water absorption and good abrasion resistance. A low water absorption is important when thin PC overlays are used in pavements and bridges because they prevent the intrusion of water and salts that could cause corrosion of the reinforcing steel in the portland cement concrete substrate. The abrasion resistance helps PC overlays against the action of traffic in pavements, floors, and bridges, and against the erosion of water in dams and other hydraulic structures.

Table 2 - *Durability Properties of PC*

Water absorption (% by weight)	< 1
Chemical resistance	
Sulfuric acid	Very good
Hydrochloric acid	Very good
Sodium hydroxide	Good
Acetone	Good to poor
Sand blast abrasion	Very good

Steel-Reinforced PC

Similar to portland cement concrete, PC has relatively low tensile strength compared to its compressive strength. Therefore, in some applications, it may be necessary to add reinforcing steel to PC to increase its strength capacity, its ductility, and its toughness. The load-deflection diagrams of a typical rectangular PC beam using a weight aggregate-to-resin ratio of 9-to-1, a shear span-to-depth ratio of 3.6, and a tensile steel reinforcement ratio of 2.1% are shown in Figure 10.

In the first stage of loading, a linear relationship exists between the load and deflection. As it would be the case with reinforced portland cement concrete, this proportional limit stage ends with the formation of a major vertical flexural crack and the resulting change in slope and decrease in stiffness of the reinforced beam. Compared to steel-reinforced portland cement concrete, PC exhibits higher strength, higher ductility, and higher deflection at failure.

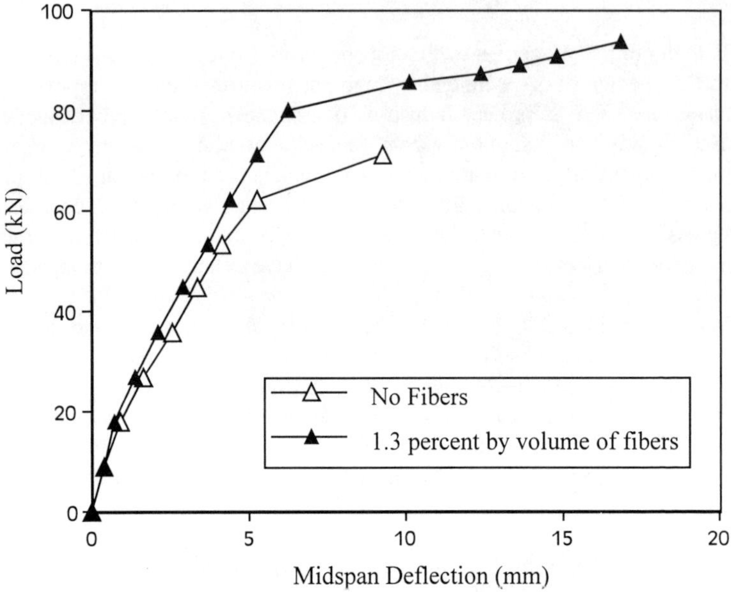

Figure 10 - *The Typical Load-Deflection Curves for Bars and Fibers Reinforced PC Beams*

Fiber reinforcements may be needed in certain applications, such as earthquake design, because they help concrete structures resist dynamic forces more efficiently by providing enough ductility and toughness. Steel fibers, with a length of 30-mm and a diameter of 0.5-mm, are added to steel-reinforced PC beams. The fibers are uncollated and hooked at their ends. Trial mixes determined that a practical limit for the steel fiber content could be no more than 2% by volume. Beyond that limit, the workability of the fresh PC mix is sharply decreased. As it can be observed in Figure 10, a small addition of 1.3% by volume of steel fibers (with low aspect ratio of 60) to a PC beam increases its ductility and toughness significantly, and to a less extent its ultimate strength. Steel fibers significantly improve the toughness and ductility of PC since the bond strength between the steel fibers and the PC is very high. Accordingly, considerable energy is required to cause a total and complete failure of the PC beams reinforced with steel fibers since the cracks could not extend without stretching, debonding or breaking the fibers.

Future Trends

Despite the fact that the use of PC is advantageous in many civil engineering applications, the transitional phase of the product from research to practice has been relatively slow. The reason for this occurrence may be attributed to the fact that the construction industry is risk averse when it comes to new technology. Indeed, the failure of a structure can have disastrous effects on its users and huge liability on its designers and constructors. Many of these failures may not be due to the use of a particular material per say; rather, they may be attributed to the incorrect use or formulation of a certain material. Therefore, it is understandable if design and construction professionals are reluctant to using a new technology when they are unfamiliar with it, particularly when codes and testing standards are not fully developed as it is the case with PC.

Another reason why PC is not yet fully accepted is because it is perceived to be much more expensive than conventional materials, particularly portland cement concrete its main competitor. Although the cost of resin used in the formulation of PC is far higher than the cost of conventional materials, the final price of finished and installed PC products can be far less than the one corresponding to conventional materials; indeed, the potential savings in labor or production costs can more than offset the high initial cost of the material, especially in applications where time is an important factor. For example, in the repair of pavements and bridges, the very fast cure time of PC permits its overnight application. Therefore, the structure can be returned to service the next morning. This property eliminates the need to construct expensive detours, improves safety, and reduces traffic congestion. In precast structures, the fast cure time of the material enables the fast and efficient use of forms and other production facilities. In addition, the good strength and durability properties of PC permit the use of thinner sections, thus reducing dead loads in structures and minimizing transportation and erection costs. Unfortunately, the final price of a finished and installed product is often ignored by design and construction professionals; rather, decisions are made solely on the material first cost alone, without considerations to design improvements, labor savings and other desirable features.

Finally, the life cycle cost of PC can be far less than conventional materials because the material is stronger and more durable than conventional material, thus requiring less maintenance and repair activities. For example, in hazardous waste containers, the very low permeability of PC to liquids and gases (unlike cement concrete, a porous material) and its inherent corrosion resistance (unlike metallic materials) allow for the safe disposal of toxic and acid wastes over long periods of time. In this case again, erroneous decisions can be made just by comparing the initial outlay investments of various alternatives, rather than accounting for the cash flow profiles of the various alternatives (initial investment and future projected cash flows) over the entire life of the structure.

It is only a question of time before PC materials is fully endorsed and accepted by the structural and construction community, especially with the development and writing of codes and standards. This endorsement will have a strong boost from government agencies if PC products are made with waste products. However, products made with waste materials will only be fully utilized if they make sense from an economical and technical point of view. PC using recycled PET and fly ash waste is an example of such a high performance, cost effective material. It is expected that future progress in physics and chemistry should allow the economical chemical conversion of other plastic wastes into resins that can be used effectively in the production of new polymer composite systems that will extend their use in civil engineering applications even further.

Conclusions

This chapter shows that PC using unsaturated polyester resins based on recycled PET and fly ash waste could be used very effectively in a variety of civil and other engineering applications such as overlays on portland cement concrete surfaces and precast components. In fact, enhanced properties are obtained with PC using resins based on recycled PET and fly ash. These properties are comparable, if not superior, to PC materials made with virgin unsaturated polyester resins. So far, the evaluation of PC using resins based on recycled PET has been confined to laboratory studies. The field application and continuous monitoring of the material would really determine the long term performance of the material under actual environmental conditions. Special precautions should be taken in cases involving large variations in temperature or large sustained loads because the viscoelastic nature of the resin binder can result in unreasonably high deformations. Special resin formulations, adequate supports, and/or large safety factors would be advisable in these instances.

Future research to improve some of the properties of the PC should focus on the development of agents that will help control shrinkage and creep. The development of new resin formulations that are insensitive to moisture effects will help simplify the use of PC materials in many of its applications. In the area of structural behavior and design of reinforced PC beams, future work should concentrate on the addition of compression and shear reinforcements, different loading and cross-sectional types, and different resin compositions and mix designs.

References

Al-Negheimish, A., (1988), *Bond Strength, Long Term Performance and Temperature Induced Stresses in Polymer Concrete-Portland Cement Concrete Composite Members*, Ph.D. Dissertation, The University of Texas, Austin, Texas.

Alkonis, J. J., Macknight, W. J., Shen, M., (1972), *Introduction to Polymer Viscoelasticity*, Wiley-Interscience, John Wiley & Sons, Inc., New York, New York.

Aniskevich, K. and Hristova, J., (1999), "Prediction of Creep of Polymer Concrete", *Journal of Applied Polymer Science*, Vol. 71, No. 12, pp. 1949-1952.

Armeniades, C. D., Dharmarjan, N., (1987), "A Constitutive Equation for Creep in Polymer Concrete," *Proceedings of the Fifth International Congress on Polymer Concrete*, Brighton, England, pp. 193-197.

Ayyar, R. S., Deshpande, S. N., (1981), "Creep Studies of Polymer Mortar, *Proceedings of the Third International Congress on Polymer Concrete*, Koriyama, Japan, pp. 504-523.

Banko, A. S., Rebeiz, K. S., Craft, A. P., (1994), "Effect of Fillers on the Temperature Relaxation Behavior of Unsaturated Polyester Based on Recycled PET", *Journal of Materials Science Letter*, Vol. 13., pp. 934-936.

Bloomfield, T. D., (1997), "Sewers and Manholes with Polymer Concrete", *Proceedings of the 1997 ASCE Conference on Trenchless Pipeline Projects*, ASCE, Reston, Virginia, pp. 466-472.

Boneti, H. J., Gunasekaran, M., Nascimento, J. and Branco, P. R., (1999), "Polymer Concrete Catenary Insulation for Mass Transit Electric Traction", *Proceedings of the 1999 IEEE/PES Transmission and Distribution Conference – A Vision for the 21st Century (v2)*, IEEE, Piscataway, NJ, pp. 670-674.

Chandra S. and Ohama Y., (1994), *Polymers in Concrete*, CRC Press, 2000 Corporate Blvd. N.W., Boca Raton, FL 33431.

Dharmarajan, N., Armeniades, C. D., (1987), "Creep Studies of Polyester Polymer Concrete Under Flexural Loading," *American Concrete Institute Special Publication 99-2*, pp. 17-30.

Doody, M. E. and Morgan, R., (1994), Polymer-Concrete Bridge Deck Overlays", *Transportation Research Record*, No. 1442, pp. 136-142.

Fowler, D. W., (1999), "Polymers in Concrete: A Vision for the 21st Century", *Cement and Concrete Composites*, Vol. 21, No. 5-6, pp. 449-452.

Guide for the Use of Polymers in Concrete (ACI 548.1R-97), (1997), ACI Committee 548, American Concrete Institute, Detroit, Michigan.

Gunasekaran, M., (1997), "Polymer Concrete: A Viable Low-Cost Material for Innovative Power Systems", *Proceedings of the IEEE International Conference on Properties and Applications of Dielectric Materials, Part 2,* IEEE, Piscataway, NJ, pp. 770-773.

Hristova, J., Bares, R. A., (1987), "Relation between Creep and Performance of PC," *Proceedings of the Fifth International Congress on Polymer Concrete*, Brighton, England, pp. 99-102.

Hsu, M., and Fowler, D. W., (1985), "Creep and Fatigue of Polymer Concrete," *American Concrete Institute Special Publication 89*, pp. 323-341.

Karcas, G. J., (1997), "Use of Polymer Concrete in Acid Containment Equipment for Electrometallurgical Plant Practice", *Proceedings of the 1997 TMS Annual Meeting*, The Minerals, Metals &Materials Society (TMS), Warrendale, PA, pp.325-332.

Kyriacou, K. C., (1991), *Accelerated Compression and Flexural Creep Testing of Polymer Concrete*, Ph.D. Dissertation, The University of Texas, Austin, TX.

Moore, G. R., Kline, D. E., (1984), *Properties and Processing of Polymers for Engineers*, Prentice-Hall, New Jersey.

Mani, P., (1983), *Some Studies on the Development and Properties of Polyester Resin-Based Polymer Concrete*, Ph.D. Dissertation, Indian Institute of Technology, Delhi, India.

Mark, A. F., (1990), *Encyclopedia of Polymer Science and Engineering, 2nd Edition*, John Wiley & Sons, Inc., 605 Third Avenue, New York, New York.

Nutt, W. O. and Staynes, B. W., (1987), "The Next Twenty Five Years," *Proceedings of the Fifth International Congress on Polymers in Concrete*, Brighton, England, pp. 413-416.

Ohama, Y., Kawakami, M. and Fukuzawa, F, (1997), *Polymers in Concrete*, E&FN Spon, 2-6 Boundary Row, London SE1 8HN, UK.

Ohama, Y., Demura, K., and Komiyama, M., (1979), "Properties of Polyester Resin Concrete with Various Styrene Contents," Proceedings, 22nd Japan Congress on Materials Research, The Society of Materials Science, Japan, pp. 360-363.

Rebeiz, K. S., Banko, A. S., Craft, A. P., (1999), "Temperature Properties of Polyester Mortar Using Fly Ash Wastes", *ASCE Hazardous, Toxic, and Radioactive Waste Management Practice Periodical*, Vol. 3, No.3, pp. 107-111.

Rebeiz, K. S., Rosett, J. W., Craft, A. P., (1996a), "Strength Properties of Polyester Mortar Using PET and Fly Ash Wastes", *ASCE Journal of Energy in Engineering*, Vol. 122, No.1, pp. 10-20.

Rebeiz, K. S., Rosett, J. W., Nesbit, S., Craft, A. P., (1996b), "Tensile Properties of Polyester Mortar Using PET and Fly Ash Wastes", *Journal of Materials Science Letter*, Vol. 15., pp. 1273-1275.

Rebeiz, K. S., Craft, A. P., (1995a), "Plastics Waste Management in Construction: Technological and Institutional Issues", *Resources, Conservation and Recycling*, Vol.15, pp.245-257.

Rebeiz, K. S., Banko, A. S., Craft, A. P., (1995b), "Thermal Properties of Polymer Mortar Using Recycled PET and Fly Ash", *ASCE Journal of Materials in Civil Engineering*, Vol. 7, No.2, pp. 129-133.

Rebeiz, K. S., Banko, A. S., Nesbit, S. M., Craft, A. P. (1995c) "Thermo-mechanical Properties of Unsaturated Polyester Resin Based on Recycled Poly(ethylene Terephthalate)", *Journal of Applied Polymer Science*, Vol. 56, PP. 757-760.

Rebeiz, K. S., (1994), "Construction Use of Polyester Composite Using Recycled PET, *Proceedings of the Third Materials Congress, Infrastructure: New Materials and Methods of Repair*, Edited by K. D. Basham, American Society of Civil Engineers, pp. 80-87.

Rebeiz, K. S., Fowler, D. W., Paul, D. R., (1993a), "High Performance Polymer Composites Using Recycled Plastics", *Trends in Polymer Science*, Vol. 1, No. 1, pp. 315-321.

Rebeiz, K. S., Fowler, D. W., Paul, D. R., (1993b), "Recycling Plastics in Polymer Concrete for Construction Applications, *ASCE Journal of Materials in Civil Engineering*, Vol. 5, No. 2, pp.237-248.

Rebeiz, K. S., Fowler, D. W., Paul, D. R., (1992), "Polymer Concrete and Polymer Mortar Using Resins Based on Recycled Poly(ethylene Terephthalate), *Journal of Applied Polymer Science*, Vol. 44, Number 9, pp. 1649-1655.

Rebeiz, K. S., Fowler, D. W., Paul, D. R., (1991), "Formulating and Evaluating an Unsaturated Polyester Composite made with Recycled PET, *Journal of Materials Education*, Vol.13, No. 5 & 6, pp. 441-454.

Rebeiz, K. S., Fowler, D. W., Paul, D. R., (1990), "Polymer Concrete Using Resins with Recycled PET", *Proceedings of the 48th Annual Technical Conference of the Society of Plastics Engineers*, Society of Plastics Engineers, Brookfield, CT, pp.1459-1461.

Vaidya, U. R., and Nadkarni, V. M., (1987), "Unsaturated Polyester Resins from Poly(ethylene terephthalate) Waste," *Industrial & Engineering Chemistry Research* , Vol. 26,.pp. 194-198.

Vipulanandan, C., Dharmarajan, N., and Ching, E., (1987), "Stress-Strain Behavior of Polymer Concrete Systems," *Proceedings of the Fifth International Congress on Polymers in Concrete,* Brighton Polytechnic, Department of Civil Engineering, Brighton, England, pp. 165-170.

Rajib B. Mallick[1]

Chapter 6

State of the Practice in Asphalt Technology

Abstract

Recent increase in traffic volume, tire pressures and cost of construction has resulted in a search for long lasting and good performing asphalt pavements. The Strategic Highway Research Program, SHRP, has resulted in a set of test methods and specifications, Superpave, for evaluation of asphalt binders and design of hot mix asphalt, HMA. The asphalt binder test methods have been developed as tools for determining the limits of asphalt binder performance temperature. The gyratory compactor applies gyratory action to simulate roller and traffic action for compacting samples of HMA. Mix equipment can be used to obtain performance-related properties. Stone Matrix Asphalt, SMA is being used increasingly as a rut resistant mix. New mixes and materials such as ultrathin mixes and polymers are being used for better performance of mixes. There is also an ongoing effort to improve mix designs for open graded friction courses and mix containing recycled materials, such as crumb rubber.

Keywords

Asphalt, HMA, Superpave, SMA

Introduction

Asphalt is one of the oldest materials known to mankind. Use of asphalt cement can be traced back to the Sumerians, who used it for shipbuilding in about 6000 BC (Barth 1962). In modern times, the first sheet asphalt pavement using naturally occurring Lake Trinidad asphalt was laid in 1876 on Pennsylvania Avenue in Washington, D.C. Today, asphalt binder, which is manufactured as a product of petroleum distillation, is present in approximately 2.3 million miles of hot mix asphalt (HMA) roadway in the United States. In 1997 about 500 million metric tons of HMA was produced in the United States.

HMA primarily consists of asphalt binder and mineral aggregate. Asphalt binder provides the bond between aggregate particles, which provide the basic skeleton for resisting loads under traffic. The performance of HMA depends on the qualities and characteristics of asphalt binder, aggregates, and the mix. Even though the basic

---

[1] Assistant Professor, Civil and Environmental Engineering Department, Worcester Polytechnic Institute, 100 Institute Road, Worcester, MA 01609.

constituents of HMA have remained the same over the years, significant changes have occurred in test methods, selection procedure of materials, and mix design systems. The American Society of Civil Engineers (ASCE) Committee on Bituminous Materials is continuously looking at new concepts in asphalt materials and technologies.

Recent trends in increase in truck traffic volume and tire pressure, high cost of pavement maintenance, and reduced budget for pavement construction and maintenance have led to the search for materials in asphalt mixes, and mix design systems that would produce longer lasting, good performing pavements. Some materials such as crumb rubber from tires and roofing shingles are also being used in HMA to reutilize waste products. While these new materials are being increasingly used in HMA, the main focus in these developments seem to be on mix design systems, which would combine the best qualities of asphalt binder, aggregate, and additives, and provide the best end product that is feasible, namely, a durable and strong HMA pavement.

The objective of this chapter is to provide a summary of currently emerging asphalt technologies. The following paragraphs describe four emerging materials/mix design systems and three emerging material additives for HMA:

1) Superpave (Superior Performing Asphalt Pavements)
2) Stone Matrix Asphalt (SMA)
3) Open Graded Friction Course (OGFC)
4) Ultrathin Mixes
5) Polymer
6) Crumb Rubber
7) Shingles

Superpave (Superior Performing Asphalt Pavements)

Superpave is the HMA mix design system that resulted from the $150 million, 5 year long Strategic Highway Research Program (SHRP) in the United States. The major objectives behind the development of Superpave were to identify and define properties of asphalt binders, aggregate, and properties of hot mix asphalt that influence pavement performance, and develop test methods for performance based specifications. Even though research continues to improve the Superpave system, at present, the Superpave system is being used by most of the state departments of transportation in the United States.

The major elements of Superpave are: (1) Superpave performance graded (PG) asphalt binder specifications, (2) aggregate properties, and (3) Superpave mix design and analysis system. Each of the elements is described in the following paragraphs.

The two major outcomes in the area of asphalt binder specifications are the introduction of performance related tests and the shift from traditional approach of testing asphalt at same temperature for different specified values of test results to testing asphalt at different temperature for the same specified value of test result. The new tests relate to three asphalt binder characteristics – temperature susceptibility, viscoelastic behavior and age hardening. The condition of the asphalt binders during the tests simulates the condition of the asphalt binder at supply, during production of HMA, and in service in the pavement. The modes of failure for which the asphalt binder characteristics are specified include rutting and cracking (temperature and fatigue related) (McGennis et al. 1994, Bahia and Anderson 1995, Anderson and Kennedy 1993).

The workability of asphalt binder is determined with rotational viscometer (RV) (Figure 1). It is used to determine the viscosity of the asphalt binder at high construction temperature (above 100°C) to ensure that the asphalt binder is sufficiently fluid for pumping and mixing. This is determined by measuring the torque required to maintain a constant rotation speed (20 RPM) of a cylindrical spindle while submerged in an asphalt binder at constant temperature. The specification sets an upper limit on the viscosity in order to achieve sufficient workability with the asphalt binder (McGennis et al. 1994).

The dynamic shear rheometer (DSR) (Figure 2) is used to characterize the viscoelastic behavior of asphalt binder, and evaluate its rutting and cracking potential. The basic principle used for DSR testing is that asphalt behaves like elastic solids at low temperatures, and as a viscous fluid at high temperatures. These behaviors can be captured by measuring the complex modulus (G*) and phase angle (δ) of an asphalt binder under a specific temperature and frequency of loading. The G* and δ parameters (Figure 3) is measured by the DSR, by applying a torque on the asphalt binder between a fixed and an oscillating plate, and measuring the resultant strain. G* is defined as the ratio of shear stress to shear strain, and δ is the time lag between the stress and the strain in the asphalt binder (McGennis et al. 1994, Christensen and Anderson, 1992). These properties are determined at high and intermediate pavement service temperatures. The high temperature is determined from the average of maximum HMA pavement temperature over a seven-day period through summer, which is obtained from data from weather stations. The intermediate testing temperature is determined from an average of seven-day average maximum and minimum design temperature (Anderson and Kennedy 1993).

In order to determine the characteristics of the asphalt binder at different times such as immediately after construction (unaged) and in-service (aged), the DSR test is conducted on both unaged, short term aged, and long term aged asphalt binder. Short term aging is done by subjecting the asphalt binder to Rolling Thin Film Oven (RTFO) test (Figure 4). The test consists of rotating a set of racks with asphalt binder bottles at a specified temperature and speed. As it rotates, the opening of each bottle passes in front of an air jet. Long term aging is simulated with the use of the pressure aging vessel (PAV) (Figure 5) The PAV test consists of subjecting the RTFO residue to high pressure and temperature inside a closed vessel. The PAV is placed inside an oven for maintaining the specified high temperature, and a cylinder of clean dry compressed air is used to supply and regulate air pressure (Bahia and Anderson 1994).

Figure 1. *Rotational Viscometer*

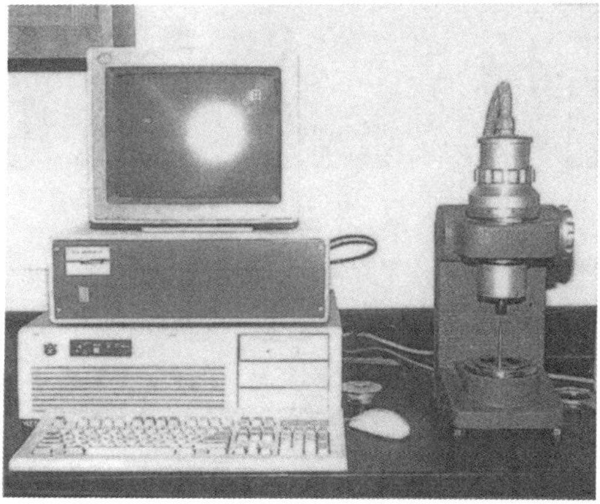

Figure 2. *Dynamic Shear Rheometer (DSR)*

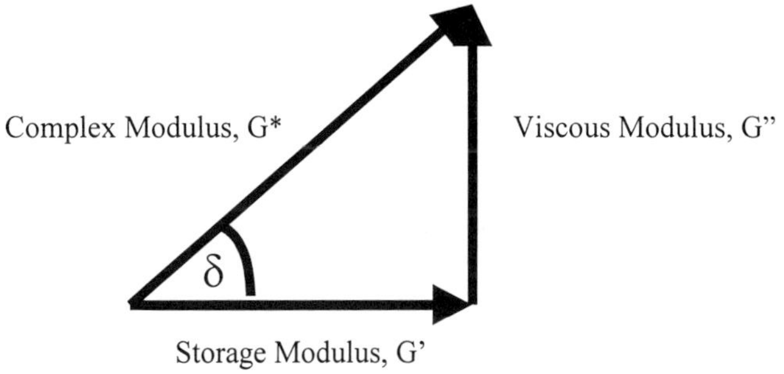

Complex Modulus is the vector sum of the storage and viscous modulus

Figure 3. *Explanation of G* and δ*

Figure 4. *Rolling Thin Film Oven (RTFO)*

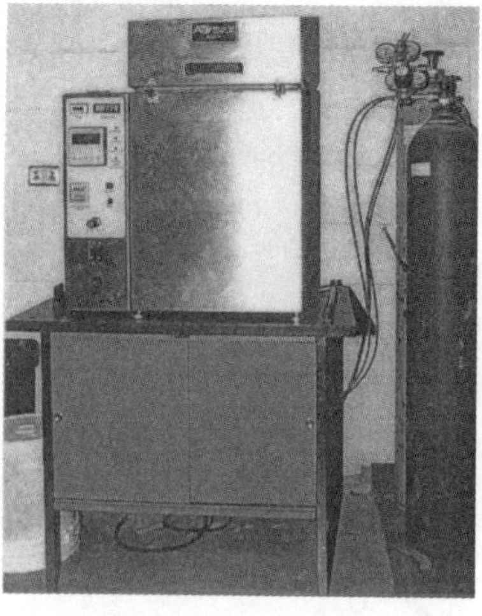

Figure 5. *Pressure Aging Vessel (PAV)*

The asphalt binder in its unaged condition, short term aged condition (RTFO conditioned) and long term aged condition (PAV conditioned) is tested with the DSR for evaluation of its rutting and fatigue characteristics. The rutting potential is evaluated by the rutting factor, defined as G*/Sin δ, of which G* and δ can be obtained from DSR testing. Since higher G* means greater resistance to rutting, and lower δ means a more elastic asphalt binder, higher G* and lower δ values are desirable to make the HMA less susceptible to rutting. Considering rutting as a stress controlled, cyclic loading phenomenon it can be shown that rutting potential is inversely proportional to G*/Sin δ. Hence, in the Superpave specifications a minimum limit is set on the value of G*/Sin δ. Testing asphalt for the rutting factor prior to long term aging is critical, since with aging the stiffness of the asphalt binder increases and resistance to rutting increases (Bahia and Anderson 1995). Hence, DSR test for rutting potential is conducted on RTFO residue. It is also conducted on unaged asphalt binder, since in some cases RTFO may not indicate the actual aging of the binder during production and placement of HMA

The asphalt binder is also tested with the DSR to determine the fatigue cracking potential of HMA. The fatigue potential is measured by the fatigue factor, defined as G* Sin δ. This is done because it can be shown that under strain controlled process, the work done for fatigue is proportional to G* Sin δ, and hence a lower value of G* Sinδ means a lower potential of fatigue cracking. This can be explained by the fact that as G* decreases, the asphalt binder becomes less stiff, and is able to deform without building up large stresses, and as δ value decreases, the asphalt binder becomes more elastic and hence can regain its original condition without dissipating work. An upper limit is set on G* Sin δ value in Superpave specification, in order to minimize the potential of fatigue cracking (Bahia and Anderson, 1995). Since the asphalt binder becomes stiffer due to aging during its service life and becomes more susceptible to cracking, the DSR test for fatigue factor determination is conducted on PAV aged samples.

The Bending Beam Rheometer (BBR) is used to evaluate the stiffness of asphalt binder, and its cracking potential at low temperature. The BBR (Figure 6) consists of a loading mechanism in which an asphalt binder beam is subjected to a transient creep load at a constant low temperature. The creep stiffness and the rate at which the stiffness changes with loading time give an indication of its cracking potential at test temperature. Higher the stiffness, and lower the rate, greater is the potential for cracking. A lower slope or rate of change in stiffness means a reduced ability of the asphalt binder to relieve thermal stress by flowing. Hence, in the Superpave specification, a maximum limit is set on the value of stiffness, and a minimum limit is set on the value of slope. Since the asphalt binder becomes stiffer with aging, and more susceptible to thermal cracking, BBR test is conducted on PAV aged samples (McGennis et al. 1994, Bahia and Anderson 1994).

In some cases, such as modified binders, a measure of stiffness is not adequate to characterize the cracking potential. This is because these binders may have a higher stiffness, but they may not be susceptible to cracking, as they are more ductile and can stretch without breaking. The direct tension test (DTT) (Figure 7) is used to characterize such binders. It is used for testing any binder which has a stiffness exceeding the specified stiffness at low temperatures. The DTT consists of a tensile testing machine, which measures the strain at failure of a dog bone shaped asphalt

Figure 6. *Bending Beam Rheometer (BBR)*

Figure 7. *Direct Tension Test (DTT)*

binder specimen, which is subjected to tensile load at a constant rate. A minimum limit is set on the strain at failure (McGennis et al. 1994).

In the Superpave system of specifications, the requirements for results from tests on asphalt binders are the same. However, the temperature at which the requirements are met is used to classify the asphalt binder. For example, a binder classified as PG 64-22 means that the binder meets high temperature physical property requirements at least up to a temperature of 64°C, and low temperature physical property requirements at least down to –22°C. There is also a provision of using higher grade asphalt (PG 76-22, for example) for mixes to be used in slow transient or standing load areas. The asphalt binder to be used in a specific project is selected on the basis of geographical location of the project, the pavement temperature and air temperature, and also on the basis of specific reliability that the user wants to have in selection of the binder (McGennis et al. 1994).

In the case of mineral aggregates, Superpave has specified criteria for use of certain test properties such as coarse aggregate angularity and flat and elongated particles content, on the basis of location of mix, and traffic volumes. The most important recommendation in Superpave for aggregates is in the case of aggregate gradation. Superpave recommends certain control points and a restricted zone in the usual 0.45 power chart that is used for plotting gradation. Control points and the restricted zone are provided to avoid the use of undesirable gradations, which can result in tender mixes. Examples of control points and restricted zone are shown in Figure 8. Superpave recommends aggregate gradation below the restricted zone, and moving of gradation closer to the control points with an increase in traffic volume (McGennis et al. 1994).

In the case of mixes, Superpave system consists of compaction of mixes with the Superpave gyratory compactor (Figure 9), and testing of mixes, based on volumetric properties as well as mechanical test properties. The Superpave gyratory compactor applies a gyratory and kneading action to the HMA sample to simulate the action of roller and traffic on HMA. The compactive effort for compaction of the HMA samples is specified in terms of the number of gyration, which is based on the volume of design traffic. The mix design and analysis system comprises of a volumetric mix design system and two levels of mix design procedure with performance testing with mechanical tests and models for prediction of low temperature and fatigue cracking, and permanent deformation potential. The volumetric mix design system, which is being currently use, consists of selection of aggregate gradation based on volumetric properties such as voids in mineral aggregate (VMA), and volumetric properties of compacted mixes, such as VMA, voids in total mix (VTM), and voids filed with asphalt (VFA). Specified values of VTM are provided for samples compacted to design gyrations and also at gyrations lower and higher than design gyrations, to avoid the production of tender and rut susceptible mixes (McGennis et al. 1994).

The two new performance testing equipment are the Indirect Tensile Tester (IDT) (Figure 10) and the Superpave Shear Tester (SST) (Figure 11). The output from these tests provides the input to performance models in Superpave to estimate actual pavement performance. Depending on the level of design (higher the required reliability, higher is the design level), different types of tests such as uniaxial strain, frequency sweep or simple shear tests can be conducted on HMA samples. Much simpler "torture" test can also be conducted for quick indication of permanent deformation potential of HMA mixes (McGennis et al. 1994).

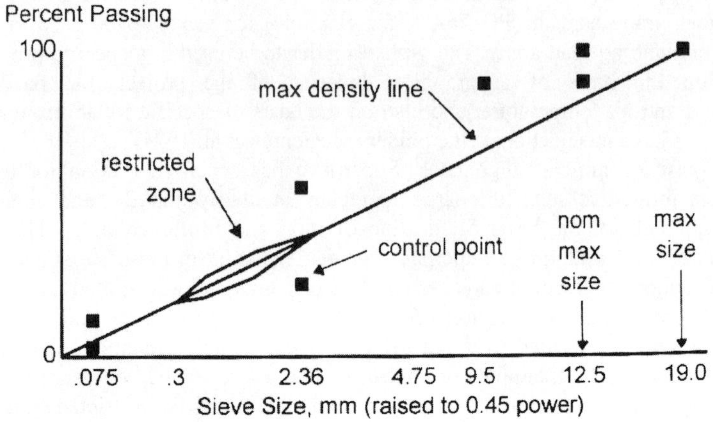

Figure 8. *Example of control points and restricted zone*

Figure 9. *Superpave Gyratory Compactor*

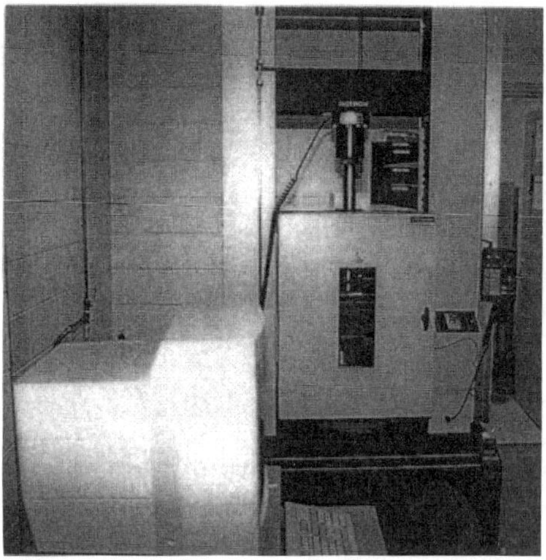
Figure 10. *Indirect Tensile Tester*

Figure 11. *Superpave Shear Tester*

At present work is continuing on the proper specification for use of modified binders, refinement of compactive effort for sample compaction, evaluation of restricted zone in aggregate gradation, and refinement of models for performance prediction in levels two and three Superpave mix design system.

Stone Matrix Asphalt

Even though Stone Matrix Asphalt (SMA) had its origin in Europe in the mid 1960s, its use in the US began in the early nineties. SMA has a gap graded structure with about 20-30 % material passing the 4.75 mm sieve, a high content of good quality crushed aggregate, and a relatively high binder content of 6-7.5 %. Stabilizers in the form of fiber and polymer are also added to SMA to avoid excessive draindown of binder from mix. A relatively high percentage (about 10 %) of material passing the 0.075 mm sieve is used in order to provide a thick and stable matrix of the binder (Kast 1985). Figure 12 shows a comparison of a schematic of a section though a HMA and a SMA mix (Brown and Mallick 1995).

The use of SMA has several advantages, which include increased resistance to rutting, increased durability, reduced vehicle noise and increased frictional resistance. The increased resistance against rutting comes primarily from the existence of stone on stone contact in the mix, as shown in Figure 12. Stone on stone contact condition can be evaluated from a plot of volumetric parameters against change in gradation. In order to achieve stone-on-stone contact condition, the criterion of 20-30% material passing the 4.75 mm sieve is specified for design of SMA (Brown and Mallick 1995a).

The increased durability of SMA comes from the thick binder film, resulting from the high binder content. The thick binder film also results in lower thermal cracking potential of mixes. The main idea behind design of SMA is to provide as much binder as possible and yet maintain a strong aggregate skeleton and avoid excessive draindown of binder from the mix. The draindown potential of the mix can be evaluated with a draindown test in the laboratory during mix design (Brown and Mallick 1995b).

The surface texture of SMA is primarily responsible for reduction in higher frequency noises. Comparison of surface texture profiles show that compared to a dense graded mix, there are deeper "valleys" in SMA, which causes a reduction of noise by as much as 2.5dB (Hoppe 1991). The surface texture also helps in obtaining a higher amount of frictional resistance. The qualities of SMA also make it possible to apply the mix in thin layers.

The steps in design of SMA basically consists of selection of proper aggregate and asphalt binder, selection of proper aggregate gradation based on two critical gradations – percentage passing the 4.75 mm and 0.075 mm sieves, selection of stabilizer, determination of optimum asphalt content (with a lower limit on asphalt content), and evaluation of draindown and moisture susceptibility of mixes (NAPA 1994).

In the United States, a study on SMA sponsored by NCHRP (Brown and Cooley 1999) has been completed. The general emphasis in the NCHRP report is on developing design and construction guidelines and specifications for the use of SMA.

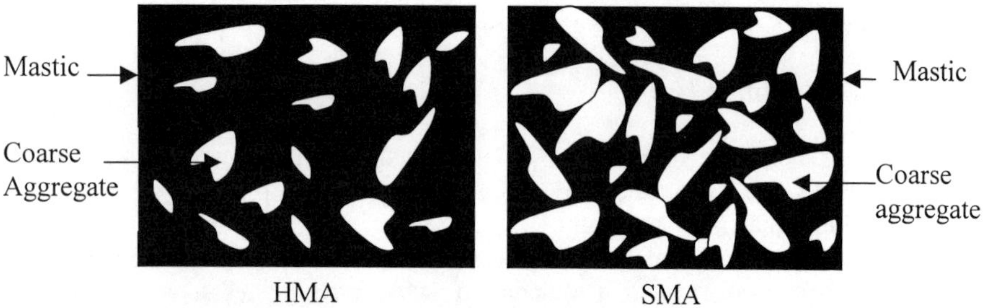

Figure 12. *Schematic of sections through HMA and SMA*

Open Graded Friction Course (OGFC)

Open graded friction courses (OGFC) are gap graded mixes with a relatively high air voids content (15-20 %), and which are typically used in very thin layers (35-40 mm) (Halstead 1983). OGFC s are being used increasingly in different countries, including the United States, primarily because of its water draining capability, improved skid resistance, and reduced traffic noise. The improved water draining capability is derived from the open aggregate structure, which is a result of very coarse gradation (Figure 13). The voids are very high, and they are interconnected, providing a means of rapid draining of water. The improved drainage of OGFC reduces splash and spray behind vehicles, eliminates reflection from the surface of wet pavements and also improves the visibility of pavement markings. OGFC mixes improves wet weather skid resistance at high speeds through improved wheel contact with the surface.

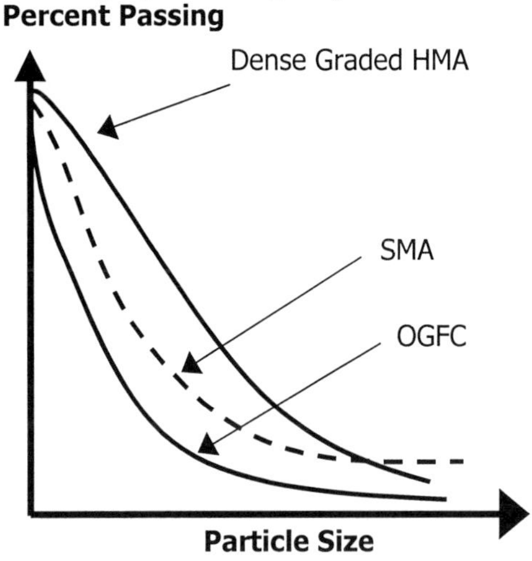

Figure 13. *Gradation of OGFC (HMA and SMA are also shown)*

The high void content, combined with layer thickness of at least 40 mm, offers the additional advantage that noise generated by vehicles is absorbed by the layer, resulting in a reduction of noise (3 to 8 dBa depending on dry or wet condition) (PIARC 1993).

The mix design system for OGFC is still evolving in different parts of the world, including the United States. Primarily, the design steps consist of determination of aggregate gradation to achieve an open structure and required permeability, selection of a relatively high binder content to improve durability and avoid loss of aggregate due to abrasion, and selection of additives to reduce draindown and/or improve durability and rutting performance. The basic principle in the design of OGFC mixes seems to be the selection of the coarsest gradation and the maximum asphalt binder that is possible, without having excessive draindown, and so as to achieve required permeability and durability.

Although OGFC have been used since the 1960s, at present, there is a renewed interest in the use of OGFC in the US, and various research are being conducted on proper mix design of OGFCs particularly with respect to adequate durability and adequate air voids for free draining of water. Studies are also being conducted in Europe to develop a cyclic test method for evaluating the potential of loss of particles from surfaces of aged OGFC mixes (Tolman and Gorkum 1996).

Ultrathin Mixes

Ultrathin mixes were first used in France in 1988, and their use has since grown in the US and other parts of the world. Ultrathin mix consists of a thin layer – 10 – 20 mm of hot mix asphalt, placed over a heavy tack coat of emulsified asphalt. These mixes improve skid resistance and water proofing ability of the surface, and are used primarily for preventive maintenance or surface rehabilitation (Chazal 1990).

The characteristics of ultrathin mixes lie both in their composition and method of construction. They are gap graded mixes, and include a large proportion (70-80%) of single sized crushed aggregate. The maximum size of aggregate most frequently used is 10 mm (Chazal 1990, Serfass et al. 1993). The typical binder content ranges from 5.2 to 6 %. Specialized equipment is used for construction of ultrathin mixes. This equipment applies the emulsified binder, and the hot mix asphalt in the form of precoated chips, in one single pass. The emulsified polymer modified binder is applied at a rate of 0.6 to 1.2 l/m^2. The mix is placed immediately after the application of the tack coat, and is compacted with a screed (Kandhal and Lockett 1997). A schematic of the equipment is shown in Figure 14. The tack coat and the mix are placed at a speed of 15 to 25 m per minute, resulting in a very speedy construction, and a reduction in traffic delay and related inconveniences.

Currently, trial sections are being evaluated at different parts of the United States to evaluate its performance under different climatic conditions. Trial mixes are being used mainly for surface rehabilitation purposes on pavements, and also as thin surfacing on bridge decks.

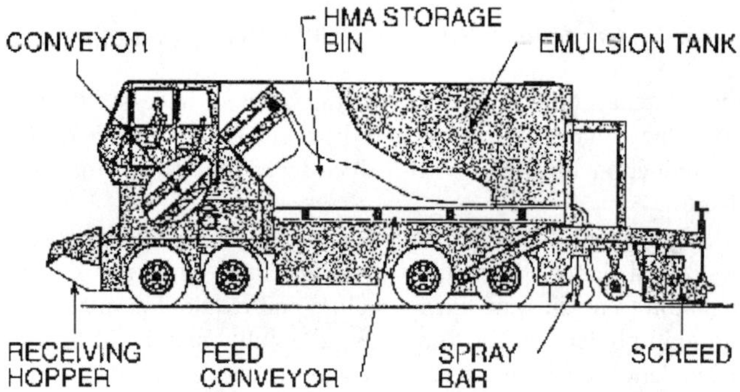

Figure 14. *Schematic of machine used in ultrathin mix construction*

Polymers

Even though the use of polymers in the US started in the1970s, polymers are only recently emerging as a commonly used additive in both Europe and the United States. This emergence is being accompanied by renewed academic interest and research in the area of polymer and polymer application in HMA.

Polymers are primarily used to improve durability of HMA pavements. The benefits of use of polymer include an increase in serviceable temperature, increase in asphalt aggregate bonding, decrease in permanent deformation potential, decrease in draindown potential, and decrease in raveling and stripping in HMA (King et al. 1986).

Two types of polymers are commonly used in HMA - elastomers and plastomers. Elastomers include block copolymers, random polymer, natural and synthetic latex. The polymers are either preblended with liquid asphalt or blended at the HMA plant. The elastomeric polymers are used primarily for decreasing temperature susceptibility, increasing strain recovery, increasing tensile strength at high strain, and increasing cohesion and adhesion properties of the asphalt binder. Styrene Butadiene Rubber (SBR) is an example of elastomeric polymer that is used in HMA. SBR can increase the stiffness of asphalt binder at long loading time (or low frequency). This improvement is significant for reducing mix tenderness and prevents rutting in HMA. At low temperatures, SBR can improve the elastic characteristics of the asphalt binder, without increasing its stiffness. This improvement can help in reducing the potential of flexural fatigue cracking in HMA.

Plastomers consist of polyethylene and ethylene copolymers. They are used primarily to decrease temperature susceptibility, increase modulus, and increase tensile strength. Ethylene Vinyl Acetate (EVA) is an example of plastomeric polymer that is used with asphalt binder. The EVA content, which can vary from a few percent to more than 50 percent, determines the mechanical properties of the asphalt binder, and the compatibility of the polymer with the asphalt binder. Generally, the properties that are modified include strength, compatibility, tenacity and stiffness. This type of polymer is used for improving workability and reduces the rutting potential of HMA (Hines 1993).

Currently, in the US a national level study is being conducted on the development of performance based specification for the use of modified asphalt binders. At present the use of polymers in asphalt binder or HMA is done according to guidelines and specifications given by the polymer manufacturer.

Crumb Rubber

Crumb rubber, obtained from recycling of vehicle tires, is emerging as a commonly used additive in asphalt binder in the United States. Started primarily as a way of recycling tires, and avoiding filling up of landfill space, the use of crumb rubber has grown steadily since the early nineties, and continues to grow as different advantages of its use are being investigated.

Generally, crumb rubber is used at about 15 to 25 percent of the total mass of the binder. It is crushed to a maximum size (generally 1 mm) before being added to asphalt binder at elevated temperature (wet process), or to aggregate at HMA plant (dry process) (Heitzman 1992). The reported benefits of using crumb rubber in asphalt mixes include greater flexibility, rougher surface texture, greater resistance to aging, and decrease in draindown potential of asphalt binder. These qualities translate to greater fatigue life, better surface friction, improved elasticity and ability to add a higher amount of asphalt binder to increase film thickness in mixes such as OGFC (Choubane et al. 1999).

At present work is continuing in different state in the United States on characterization of crumb rubber modified mixes, finding more application areas of crumb rubber, and proper design of crumb rubber modified HMA.

Shingles

One form of reclaimed materials being used increasingly used in the United States is roofing shingles. Shingles are primarily composed of asphalt cement, mineral aggregate, fiberglass and cellulose, with a wide variation in the relative proportion of these materials. The use of shingle started as a way of recycling, and avoiding waste of the material. Since shingle contain about 20 to 30 % asphalt cement, its use in HMA can reduce the need for new asphalt binder significantly, and thus result in substantial savings (Paulsen et al. 1988).

Generally shingles are shredded down to a size of 12 mm or less prior to introduction in the mix to ensure melt down and uniform dispersion the mix (Button et al. 1995). Reported advantages of using shingles include decrease in optimum asphalt binder content and thus reduction in cost, decrease in freeze thaw potential of HMA, decrease in low temperature stiffness, and decrease in rutting potential of HMA.

Currently research is being conducted on the use of shingles in HMA. A significant amount of shingles has been used in several paving projects, including one in the construction of parking lot in Disney World in Florida.

References

Anderson D.A. and Kennedy, T. W., (1993), "Development of SHRP Binder Specifications", *Journal of AAPT*, Vol. 62, 1993.

Bahia H.U. and Anderson, D.A., "The Pressure Aging Vessel (PAV): A Test to Simulate Rheological Changes Due to Field Aging", *ASTM Special Technical Publication* 1241, 1994.

Bahia H.U. and Anderson, D.A., "The SHRP Binder Rheological Parameters: Why Are They Required and How Do They Compare to Conventional Properties", *Preprint paper* No. 950793, Presented at the Transportation Research Board Meeting in Washington, D.C, 1995

Barth, E.J. *Asphalt Science and Technology*, Gordon and Breach Science Publishers, New York, NY, 1962.

Brown E.R and Mallick, Rajib. B. "An Evaluation of Stone on Stone Contact in Stone Matrix Asphalt", *Transportation Research Record* 1492, Transportation Research Board, National Research Council, Washington, D. C., 1995.

Brown E.R and Mallick. Rajib B., "A Laboratory Study on Draindown of Asphalt Cement in Stone Matrix Asphalt (SMA)", *Transportation Research Record* Transportation Research Record No. 1513, Transportation Research Board, National Research Council, Washington, D. C., 1995.

Brown E.R and L.A. Cooley, Jr. *Developing Stone Matrix Asphalt Mixtures for Rut-Resistant Pavements.* Project D-09-08. NCHRP Report 425. Transportation Research Board, National Center for Asphalt Technology, National Research Council, Washington, DC, 77 p, 1999, ISBN 0309063191.

Button J et al. *Shingles and Toner in Asphalt Pavements.* FHWA Research Report FHWA/TX-96/1344-2F. Texas Transportation Institute, 1995.

Chazal, Pierre. *NOVACHIP.* First International Conference in Search of Performance Excellence, Hawaii, 1990.

Choubane, Bouzid., Sholar, Gregory, A., Musselman, James. A., and Page, Gale. C., "*A ten-year Performance Evaluation of Asphalt-Rubber Surface Mixes*", *Preprint paper* No. 00926, Presented at the Transportation Research Board Meeting in Washington, D.C, 1999.

Christensen, D.W. and Anderson, D. A., "*Interpretation of Dynamic Mechanical Test Data for Paving Grade Asphalt Cement*", Journal of AAPT, Vol. 61, 1992.

Halstead, W.J. *Criteria for Use of Asphalt Friction Surfaces*, Transportation Research Board, NCHRP Synthesis104, 1983.

Hines. M.L. *Asphalt Cement Properties Improved by Styrelf – Laboratory and Field Data*, Koch Materials Company, 1993.

Heitzman M.S. *State of the Practice- Design and Construction Of Asphalt Paving Materials with Crumb Rubber Modifier.* Federal Highway Administration Report Number No. FHWA –SA-92-022, 1992.

Hoppe, W.. *Splittmastixasphalt – A construction Method with many Applications*, Bitumen 2/91, 1991.

Jean-Pierre Serfass, Bense, Pierre, and Samanos, Jacques, "Performance Assessment of Ultrathin Friction Courses", *Preprint Paper* No. 930541. Presented at the Transportation Research Board Meeting in Washington, D.C, 1993.

Kandhal, P.S. and Lockett, L., *Construction and Performance of Ultrathin Asphalt Friction Course*, NCAT Report 97-5, Auburn University, Alabama, 1997.

Kast O.E. *Long Term Experience with Splittmastixasphalt in the Federal Republic of Germany*, Eurobitume, 1985.

King, Gayle N.; Muncy, Harold W. and Prudhomme, Jean B., "*Polymer modification: binder effect on mix properties*", Proceedings of the AAPT, Volume 55, 1986.

McGennis, R.B., Shuler, S., and Bahia, H. U., *Background of Superpave Asphalt Binder Test Methods*. FHWA Report No. FHWA-SA-94-069, 1994.

National Asphalt Pavement Association. *Guidelines for Materials, Production, and Placement of Stone Matrix Asphalt (SMA)*, IS 118, 1994.

Paulsen G., Stroup-Gardiner, M. and Epps, J. A., *Roofing Waste in Asphalt paving Mixtures*. Center for Construction Materials Research, University of Nevada at Reno 1988.

PIARC Technical Committee on Flexible Roads and on Surface Characteristics, *Porous Asphalt* Publication No. 08.01B , PIARC, Paris, 1993.

Tolman, F, and Gorkum, F., "Mechanical Durability of Porous Asphalt, Paper presented at Eurobitume", 1996.

Richard E. Klingner[1]

Chapter 7

Emerging Masonry Materials

Abstract

In this chapter, emerging developments in masonry materials, structural systems, construction techniques, and evaluation techniques are viewed in the overall context of the masonry technical community, and are discussed in more detail. Emerging masonry materials include low-density conventional units, autoclaved aerated concrete units, mortar cement, and flashing materials. Applications include new structural systems (such as prestressed masonry), evaluation of existing buildings, and new test methods for investigating flexural bond strength. Relationships between emerging masonry materials and their applications are summarized in Table 1 and discussed in the text. Finally, future trends in masonry materials and applications are predicted, and the corresponding research needs are projected.

Keywords

Design, innovation, masonry, structures

Introduction

Masonry involves four basic components: units; mortar, grout; and accessories. Masonry units are most commonly made of concrete or fired clay. Masonry mortar holds units together, and also holds them apart, in the sense that the variable thickness of conventional mortar joints can compensate for dimensional tolerances in masonry units. Masonry grout, essentially fluid concrete with or without pea-gravel aggregate, is used to integrate masonry structurally and surround deformed-bar reinforcement. Accessories include reinforcement, connectors, and waterproofing materials.

Masonry, while the world's oldest construction system, is currently undergoing changes in many areas. Some changes involve individual components or materials: familiar and new types of masonry units; new types of masonry mortar and ways of preparing it; and new connecting and waterproofing materials. Other changes involve masonry structural systems or construction techniques. Still other changes involve ways in which masonry is evaluated.

---

[1] Phil M. Ferguson Professor in Civil Engineering, Department of Civil Engineering, University of Texas, Austin, Texas 78712

Table 1 Interrelationship of changes in masonry materials, design, construction and evaluation

Materials	Structural Systems	Construction Techniques	Evaluation Approaches
Units • low-density conventional units • AAC units			
Mortar • mortar-cement mortar		• silo-mix mortar	
Grout		• grouting admixtures	
Accessory Materials • integral water-repellent admixtures for units and mortar • modern flashing materials			
Masonry in General	• autoclaved aerated concrete masonry • prestressed masonry		• serviceability • existing buildings • bond strength • "green" masonry

The interrelationship among some of these changes is shown in Table 1. In that table, the essential materials of masonry (units, mortar, grout and accessory materials) are shown in the rows; applications of those materials (structural systems, construction techniques, and evaluation approaches) are shown in the columns. Intersections of rows and columns show ways in which emerging materials are related to new applications. For example, under the row (material) entitled "Masonry in General," and the column (application) entitled "Structural Systems," Table 1 notes "autoclaved aerated concrete masonry" and "prestressed masonry." Those subjects are discussed in this chapter. Many row-column intersections in Table 1 are blank, indicating only that no topics related to those intersections are discussed here. One useful way to approach Table 1 is to examine each topic treated in this chapter, see the row-column intersection on which that topic lies, and observe the materials and applications affected by that topic. For example, autoclaved aerated concrete masonry involves advances in materials, and also possible advances in structural systems. Finally, the table can provide insight into the general relationships between emerging masonry materials and their applications.

In the remainder of this chapter, each type of change is discussed, with specific examples. Finally, future trends and research needs are projected.

Emerging Trends in Masonry Materials

Conventional Masonry Units

Most conventional masonry units used in the United States are of fired clay (ASTM C62 or C216), or lightweight concrete (ASTM C90). The former typically have a density of about 2000 kg/m^3 (120 lbs/ft^3), and compressive strengths from 56 to 200 MPa (8,000 to 30,000 lb/in^2.). The latter typically have a density of 1500 to 1700 kg/m^3 (90 to 105 lbs/ft^3), and compressive strengths from 13 to 20 MPa (1900 to 3000 lb/in.^{2}. In recent years, considerable effort has been devoted to the development of masonry units that are lighter than this, and that still have acceptable structural characteristics. There are several reasons for this: first, lighter units use less material, and thus have the potential to be made at lower cost; second, lighter units can be laid with less effort, thereby lowering labor costs and reducing time lost due to injuries; and third, lighter units are often more effective thermal insulators, and may consequently produce a finished structure with lower energy costs.

Low-density clay masonry units are most often produced by using clay mixed with polystyrene or hollow ceramic beads. The former vaporize in the firing process. This type of masonry is widely used in Europe (Righetti and Calliari 1989), but almost unused in the United States.

Low-density concrete masonry units are normally produced using low-density aggregate and zero-slump concrete. Further decreases in density can be achieved by reducing the dimensions of the units themselves (Bowser *et al.* 1996), or by substituting other materials for some of the aggregates (Goyal *et al.* 1994).

Autoclaved Aerated Concrete Masonry Units

In addition to the efforts involving the masonry units discussed above, extensive efforts are now underway to open the US masonry market to a material that has been used for more than 60 years in Europe, but is almost unknown in this country: autoclaved aerated concrete, or AAC (RILEM 1993, Nelson and Barnett 1997). AAC is a lightweight calcium silicate hydrate, composed of portland cement, finely ground silica sand, high-calcium quicklime and water, plus a small amount of aluminum powder. When these ingredients are combined, the slaking of the quicklime heats the mixture sufficiently to induce initial set in the cement, and the aluminum powder reacts with the basic cement to produce tiny bubbles of hydrogen gas, which make the mixture rise. In this state, the "cake," while still soft, has sufficient strength to be cut or formed. The cut or formed cake is then autoclaved under pressure, completing the hydration process.

The resulting calcium silicate hydrate has very low density (as low as 200 kg/m^3), with excellent acoustical characteristics, good thermal insulation characteristics, and good thermal mass. It must be protected from the elements (for example, with a coating of stone, stucco, or even paint). It is strong enough for many structural applications. It can be cut and formed with conventional woodworking tools.

The strength of AAC masonry units depends on the density of the concrete used to make them. AAC is often characterized as low-strength, medium-strength, or high-strength. Low-strength AAC has a density of 200 to 400 kg/m^3, and a compressive strength of less than 1.8 MPa. Medium-strength AAC has a density of 300 to 600 kg/m^3, and a compressive strength of 1.8 to 4.0 MPa. High-strength AAC has a density of 500 to 1000 kg/m^3, and a compressive strength of 4.0 MPa or more (RILEM 1993).

Mortar Cement

The two traditional masonry mortars (portland cement-lime mortar and masonry-cement mortar) have recently been joined by a new product, "mortar-cement mortar." This new mortar is made by mixing water and masons' sand with a new product, "mortar cement" (ASTM C1329). Mortar cement is a proprietary product generally similar to masonry cement; it differs primarily in its air content, which is generally lower than that of masonry cement. It has a minimum specified flexural bond strength of about 0.7 MPa (100 $lb/in.^2$) (Melander and Ghosh 1995, Melander 1996), generally comparable with that obtained with portland cement - lime mortars (Hedstrom *et al.* 1991). Mortar-cement mortar is intended to compete with portland cement-lime mortar in applications where high bond strength is important.

Grouting Admixtures

Reinforced masonry, widely used in parts of the US where high winds or strong earthquakes are design concerns, uses clay or concrete masonry units, with deformed reinforcement placed in grout, a fluid concrete mixture (Klingner 1995). Masonry grout is usually specified in accordance with the proportion specifications of ASTM C476 (1 volume portland cement, 3 volumes sand, 2 volumes pea gravel, and enough water to produce a slump of 200 to 275 mm (8 to 11 in.). Research conducted as part of the TCCMAR program (Noland 1990) showed that void-free grouting was practically impossible to achieve without grouting admixtures, and reasonably easy to achieve with such admixtures. Grouting admixtures typically contain a mixture of superplasticizers, water-retention agents, and shrinkage compensating agents (Hooker 1992).

Masonry Accessory Materials

Although innovations in masonry unit, mortar, and grout materials are the most obvious examples of material innovation in the masonry industry, it is important not to underestimate the combined effects of innovations in materials intended to produce more durable, longer-lasting masonry. These include: integral water-repellent admixtures; higher-quality, easier-to-use flashing materials; and more durable connectors.

Integral water-repellent admixtures are used with concrete masonry. They replace water-repellent surface coatings. The most common integral water repellents are polymers and stearates. All are added to the plastic concrete when it is mixed. Because they tend to reduce the absorption of the concrete masonry units, these admixtures can reduce flexural bond strength. For that reason, when integral water-repellent admixtures are used in the units, compatible mortar admixtures must be used in the mortar. These are usually polymer admixtures that increase bond strength (Schierhorn 1994).

Flashing is used in masonry construction to re-direct water back to the outside after it has penetrated the masonry. Polyvinyl chloride (PVC) is the most common flashing material. However, it degrades rapidly, especially under UV exposure. Copper and stainless steel have traditionally been used when durability was an important objective. However, these materials are expensive, and difficult to install correctly. In recent years, flashing systems have been developed that are easier to install than metal, and much more durable than thin PVC. These include self-adhering asphaltic or EPDM membranes, sometimes with metal edges in areas of UV exposure; plastic-coated fiberglass; and heavy-gage PVC (Schierhorn 1994). Examples of self-adhering flashing installation are shown in Figure 1.

Figure 1 Example of masonry flashing

Emerging Trends in Masonry Structural Systems

Autoclaved Aerated Concrete Masonry

Autoclaved aerated concrete masonry can be used in the form of large panels, or masonry units. The large panels are typically reinforced with wire cages. AAC masonry units are generally unreinforced. They are laid with thin joints, using modified cementitious mortar. AAC masonry units are available in thicknesses from 200 to 300 mm, in heights of about 300 mm, and in lengths up to 600 mm. Examples of AAC masonry construction are shown in Figure 2. As shown in that figure, large AAC panels can be used for wall, roof and floor elements. AAC masonry elements can be used for walls.

Figure 2 Possible structural application of AAC panels and masonry units

Prestressed Masonry

Prestressing has been used worldwide for concrete structures for many years. It has been used less widely in Europe for masonry structures (Shaw 1996, Ganz 1996, Ganz and Shaw 1997). It has been used in some parts of the United States (Schultz and Scolforo 1991). Prestressing is expected to be most useful in increasing the out-of-plane flexural strength of slender masonry walls, permitting masonry to be used advantageously in buildings now constructed primarily of tilt-up or cast-in-place concrete. Based on field and laboratory experience in this country and elsewhere (Shaw 1996), design provisions have been approved for the Masonry Standards Joint Committee Code (MSJC 1999).

The provisions include specifications for materials, and design provisions for unbonded and bonded prestressing. The design provisions include computations for the depth of the compressive stress block, general guidance for computation of capacity under combinations of axial load and flexure, computations of shear capacity, and computations of tendon development.

Emerging Trends in Masonry Construction Techniques

Methods for Preparing Masonry Mortar

In the United States, most mortar is either portland cement-lime mortar, or masonry cement mortar (Klingner 1995). While these mortar types themselves have not changed much, processes for preparing job-site mortar have evolved significantly. Two important innovations in this area are single-bag portland cement-lime mortar, and silo-mixed mortar.

Portland cement-lime mortar (ASTM C270) has traditionally been mixed by volume proportions at the job site, first using shovel counts, and then using cubic-foot measures. This process is labor-intensive and subject to error, since it involves job-site mixing of two cementitious materials (portland cement and hydrated masons' lime), plus masons' sand. Portland cement-lime mortar is easier to proportion using pre-blended mixtures of portland cement and lime, often referred to as "single-bag" portland cement-lime mortar. Although the manufacture and distribution of such "single-bag" portland cement-lime mortars is not new, it has increased in popularity in recent years.

On-site silo-mixing of both portland cement-lime and masonry cement mortars has also become much more widely used. In this system, a large silo mixer, containing the ingredients needed for masonry mortar, is delivered to the job site by truck. It batches and mixes mortar accurately and consistently; when it empties, it is refilled or replaced. It has many potential advantages over job-site mixing: it keeps mortar ingredients clean and dry; it takes up less area than bagged materials; it eliminates problems with theft or spoilage of bagged masonry ingredients; it eliminates the labor costs associated with batching and mixing mortar on-site; it increases mason productivity by maintaining a constant supply of consistent mortar; and if is more accurate and consistent than hand mixing (Frohmader 1991).

An example of a silo mixer is shown in Figure 3. The mixer controls the volume proportions of dry ingredients. The volume proportion of water is adjusted to produce mortar of suitable workability for the weather conditions, mortar type, and unit absorption.

Figure 3 Silo for mixing and dispensing masonry mortar

Bond Strength of Masonry

Over the past decade, considerable advances have been made in understanding the factors contributing to masonry bond strength. Those advances were primarily motivated by the desire to specify more rational allowable tensile bond stresses for unreinforced masonry, and by the desire to develop more rational criteria for deciding which types of masonry products should be permitted in different seismic hazard zones.

Research was critical to this process. An important first was the development in the early 1980's of ASTM C1072, a standardization of the bond wrench, a test method in which the bed joints of a prism are successively broken by an eccentric axial load. Figure 4 shows an example of a bond-wrench testing machine. The prism is clamped by the lower part of the machine; the topmost unit in the prism is clamped by a loading head; that head is loaded eccentrically in compression until the bed joint fails in flexural tension. Bond-wrench testing is a relatively quick and cost-effective way to obtain bond strength data.

Figure 4 Example of bond-wrench testing machine

Using bond-wrench machines and a strict testing protocol, the typical range of bond strength of portland cement-lime and masonry cement mortars was described, and techniques were developed for reducing the variability of bond strength results due to testing procedures (Hedstrom *et al.* 1991, Melander *et al.* 1993). Those results were used to set allowable bond stress values in the Masonry Standards Joint Committee Code (MSJC 1999), starting with the 1995 edition of that code. Those testing techniques have now been standardized in ASTM C1329 and ASTM C1357.

Emerging Trends in Evaluation Of Masonry

Improved Serviceability of Masonry

Along with advances in structural systems and design approaches for masonry, attention has also been paid to ways to increase the performance of masonry in service. One important aspect of this is the resistance of masonry walls to water penetration. Water can pass through a barrier wall, or through the outer wythe of a cavity wall, by any or all of the following mechanisms:

- through the units (and grout, if any) of the outer wythe;
- through the mortar of the outer wythe;
- through cracks between the units and mortar of the outer wythe;
- through weepholes in the outer wythe; or
- through gaps in the flashing protecting the outer wythe.

Generally, much more water passes through cracks at the unit-mortar interface, than through the units or mortar themselves (Birkeland 1963, Grimm 1982).

If the wall is a cavity wall, once moisture has passed through the outer wythe, and through whatever vapor-retarding barrier may be present, it can pass through the inner wythe by any or all of the following mechanisms:

- through the units (and grout, if any) of the inner wythe;
- through the mortar of the inner wythe;
- along cracks between the units and mortar of the inner wythe; or
- through gaps in the flashing protecting the outer wythe.

Clearly, the water penetration resistance of masonry walls depends on many factors, including materials, design, detailing and workmanship. Through technical publications and reference materials, all segments of the masonry industry are attempting to educator masonry construction professionals on how to build masonry walls that resist water penetration.

Connectors are used to attach masonry wythes to each other, to attach masonry to other structural elements, and to attach items to masonry. Examination of existing structures with water-penetration problems often reveals corrosion damage to connectors. This damage, if unchecked, can lead to structural failure. For this reason, the use of more-durable connectors implies significant increases in the useful life of the masonry. The useful life of electroplated steel connectors is usually about 7 years; hot-dipped galvanized connectors last longer; and stainless-steel connectors last the longest of all. In Europe and Canada, stainless-steel connectors are quite common. In the US, hot-dipped galvanized connectors are more often found, but stainless steel is coming into more frequent use (MDG 1997). Examples of connectors, taken from the Brick Industry Association Technical Notes, are shown in Figure 5.

Masonry Life Extension

While the general topic of masonry life extension is too broad for this chapter, it is useful to mention some applications of new and emerging materials for masonry repair and rehabilitation. One effective technique for masonry repair is to fill voids in the masonry and improve its overall structural strength by injecting cementitious grout (Manzouri *et al.* 1996). Such grout is usually composed of portland cement, water, grouting admixtures, and sometimes fine silica sand (Atkinson and Schuller 1993).

Another effective technique has been the use of carbon fibers bonded to the masonry with adhesive (Priestley and Seible 1995).

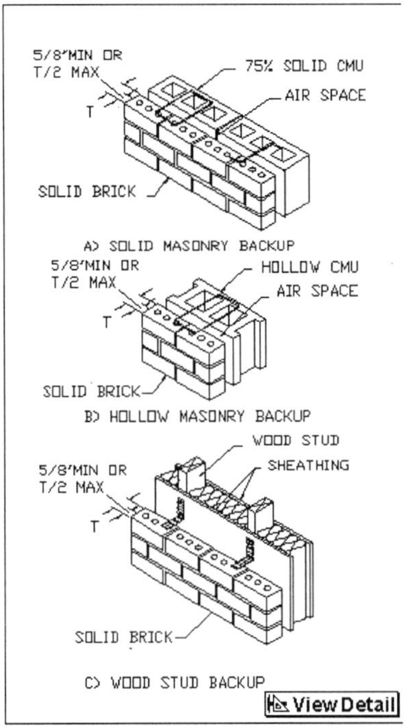

Figure 5 Examples of masonry connectors (from BIA Technical Notes)

"Green" Masonry

Over the past decade, the overall term "green" masonry has come to denote a wide range of masonry building applications and practices whose ultimate objective is the more efficient use of resources:

- "Green" masonry can refer to the use of masonry buildings with low environmental costs throughout their service life.

- "Green" masonry can also refer to efforts to reduce the environmental costs associated with the manufacture of masonry units.

The first definition, discussed by Beall (1994), includes efforts to produce masonry buildings with extended life cycles, low maintenance costs, low energy costs, low maintenance costs, healthy indoor environments, and the ability to be recycled.

Properly designed, detailed and constructed masonry can be very durable -- witness the continued usefulness of masonry structures thousands of years old. Masonry does not burn, and is durable without coatings or paints. The thermal mass of masonry can be used with passive solar design techniques to produce structures with very efficient energy use. Unlike many other building materials, masonry emits no volatile organic compounds or chemical pollutants. Finally masonry materials can either be re-used, or can be disposed of as safe, inert landfill materials.

The second definition, discussed by Hooker (1994) and Koch (1994), emphasizes efforts to reduce the environmental cost of producing masonry materials. This practice is often referred to as "sustainability." One measure of sustainability is the total inherent, or embodied, energy required to produce a building component, such as masonry. This includes the energy involved in mining the raw material for the component, in fabricating the component, transporting it to the building site, and erecting it there. Masonry components are generally considered very sustainable:

- Clay brick are made from clay or shale, finely ground, mixed with water, shaped, dried, and fired. Raw clay and shale are abundant, and cheaply transported over the short distances required. The most significant energy cost is associated with the firing process. Most kilns now use natural gas, which leads to reduced emissions and a more precise control of fuel consumption. Culled units are recycled back into the production process (Hooker 1994, Hahn 1994).

- Concrete masonry units are made from portland cement, aggregates and water. The aggregates are generally readily available. Portland cement requires considerable heat energy for its manufacture. Nevertheless, these energy costs can be offset by the partial substitution of recycled materials such as fly ash (Hooker 1994, Huizer and Day 1994), and by the use of cement kilns to burn hazardous wastes.

Efforts to enhance the overall sustainability of masonry construction are exemplified by studies of construction recycling (Hansen 1994, Speare 1995), of the ecological impact of clay masonry manufacture (Koch 1994), and by studies of re-use of masonry components in historic preservation (Slaton *et al.* 1996). In addition to those overall efforts, specific efforts have been made to incorporate waste materials such as fly ash (Huizer and Day 1994, Wei *et al.* 1994), sawdust (Goyal *et al.* 1994), glass (Huizer and Day 1994, Meyer *et al.* 1996), or dried sewage sludge (Ali and Chang 1994, Tay and Show 1994) into masonry units. Fly ash is sometimes used in autoclaved aerated concrete as well.

Outlook and Future Research Needs

In assessing the probable future trends for new materials, systems and techniques in masonry, it is important to note that advances are possible in all of these. It is also important, however, to note that advances will only occur if they offer short- and long-term economic advantages to those who use masonry as a construction system.

Outlook for Emerging Masonry Materials

At the present time (May 2000), the masonry industry in the United States is selling every unit it can produce. Under these circumstances, the masonry industry cannot be reasonably be expected to invest significantly in research on masonry materials. Exceptions include means of manufacturing conventional masonry units using significantly less raw material, significantly less energy, or significantly increased amounts of lower-cost (that is, recycled) materials. Exceptions also include producers of entirely new types of masonry units (such as autoclaved aerated concrete masonry units). The masonry industry can be expected to continue to invest in materials that increase productivity, such as improved self-adhesive flashing.

Outlook for Emerging Masonry Structural Systems

Provided that existing construction markets are able to absorb the production capacity of existing masonry producers, the masonry industry is not likely to invest significantly in research on new masonry structural systems. Exceptions include producers of accessory materials that are integral to a new system (such as prestressed masonry) or producers of entirely new types of masonry units and panels (such as autoclaved aerated concrete masonry units).

Outlook for Emerging Masonry Construction Techniques

The development of more cost-effective production techniques, such as silo-mixed mortar and self-erecting scaffolding, have made masonry construction more competitive. Research in these areas is limited, however, by lack of organized communication between mason contractors and construction engineering researchers.

Outlook for Emerging Masonry Evaluation Techniques

Because of increasing interest in life extension of existing buildings, emerging masonry evaluation techniques will be the subject of significant research in the near future. Examples include non-destructive techniques for evaluating strength; cost-effective repair techniques; and the development of building code provisions tied more closely to structural longevity. Research funding in these areas is driven by a combination of owner interest, and forensic investigation related to buildings with problems. The principal challenge in this area will be to extract life-extension data, and evaluation and retrofitting techniques, from the forensic sector into the public domain.

References

Ali, M.T. and Chang, W. F. (1994), "Strength Properties of Cement-Stabilized Municipal Solid Waste Incinerator Ash Masonry Bricks," *ACI Materials Journal*, American Concrete Institute, Detroit, Michigan, vol. 91, no. 3, pp. 256-263.

Atkinson, R. H. and Schuller, M. P. (1993), "Evaluation of Injectable Cementitious Grouts for Repair and Retrofit of Masonry," Special Technical Publication 1180, *Proceedings* of the Symposium on Masonry: Design and Construction, Problems and Repair, ASTM, Philadelphia, Pennsylvania, pp. 355-368.

Beall, C. (1994), "Masonry Buildings and Environmental Concerns," *Magazine of Masonry Construction*, Aberdeen Group, Addison, Illinois, vol. 7, no. 4, pp. 162-166.

Birkeland, O. (1963), "Rain Penetration Investigations: A Summary of the Findings of CIB Working Commission on Rain Penetration," Norwegian Building Research Institute, *Report No. 36*.

Bowser, J. D., Krause, G. L. and Tadros, M. K. (1996), "Freeze-thaw Durability of High-Performance Concrete Masonry Units," *ACI Materials Journal*, American Concrete Institute, Farmington Hills, Michigan, vol. 93, no. 4, pp. 386-394.

Frohmader, D. L. (1991), "Silo Mortar Mixers," *Magazine of Masonry Construction*, Aberdeen Group, Addison, Illinois, vol. 4, no. 1, pp. 7-10.

Ganz, H. R. (1996), "VSL's Experience with Post-Tensioned Masonry," *Proceedings*, Structures Congress - Worldwide Advances in Structural Concrete and Masonry, April 15-18, Chicago, Illinois, American Society of Civil Engineers, New York, New York, pp. 25-36.

Ganz, H. R. and Shaw, G. (1997), "Stressing Masonry's Future," *Civil Engineering*, American Society of Civil Engineers, New York, New York, vol. 67, no. 1, pp. 42-45.

Goyal, A., Rashwan, M. S., Hatzinikolas, M. A. and Zervos, S. (1994), "Structural Performance of Cavity Walls Constructed with Units Containing Sawdust and Shear-connected to the Brick Veneer," *Canadian Journal of Civil Engineering*, vol. 21, no. 4, pp. 576-584.

Grimm, C. T. (1982), "Water Permeance of Masonry Walls: A Review of the Literature," *Masonry: Materials Properties and Performance,* ASTM STP 778, J.G. Borchelt, Ed, American Society for Testing and Materials, pp. 178-199.

Hahn, P. (1994): Hahn, Peter, "Environmental Marketing for the Brick and Tile Industry," *Ziegelindustrie International*, vol. 47, no. 1, pp. 46-49.

Hansen, H. (1994), "Baustoffrecycling: Vorstellung des daenischen Recyclinghauses" ("Building materials Recycling - Concept of the Danish Recycling House"), *Ziegelindustrie International*, vol. 47, no. 1, pp. 43-45.

Hedstrom, E. G., Tarhini, K. M., Thomas, R. D., Dubovoy, V. S., Klingner, R. E. and Cook, R. A. (1991), "Flexural Bond Strength of Concrete Masonry Prisms Using Portland Cement and Hydrated Lime Mortars," *The Masonry Society Journal*, Vol. 9, No. 2.

Hooker, K. A. (1992), "Admixtures for Masonry Grout," *Magazine of Masonry Construction*, Aberdeen Group, Addison, Illinois, vol. 5, no. 6, pp. 221-223.

Hooker, K. A. (1994), "Masonry Materials and Environmental Concerns," *Magazine of Masonry Construction*, Aberdeen Group, Addison, Illinois, vol. 7, no. 4, pp. 167-171.

Huizer, A. and Day, R. L. (1994), "Environmentally Friendly Masonry Unit Manufactured from Waste Products, *International Journal for Housing Science and Its Applications*, International Association for Housing Science, Coral Gables, Florida, vol. 18, no. 4, pp. 251-262.

Klingner, R. E. (1995), "Design, Analysis and Behavior of Masonry Structures under Gravity and Lateral Loads," *Proceedings*, Third National Conference in Concrete and Masonry Engineering, San Francisco, California, June 14-17, 1995.

Koch, G. (1994), "Die Okobilanz des Ziegels" ("Bricks - An Ecological Assessment"), *Keramische Zeitschrift*, Schmid GmbH, Freiburg, Germany, v. 46, no. 2, pp. 80-82.

Krause, G. L., Devalapura, R. and Tadros, M. K. (1996), "Testing of Prestressed Clay-brick Walls," *Proceedings* of the Structures Congress on Worldwide Advances in Structural Concrete and Masonry, American Society of Civil Engineers, New York, New York, April 15-18, Chicago, Illinois, pp. 37-48.

Manzouri, T., Schuller, M. P., Shing, P. B. and Amadei, B. (1996), "Repair and Retrofit of Unreinforced Masonry Structures," *Earthquake Spectra*, Earthquake Engineering Research Institute, Oakland, California, vol. 12, no. 4, pp. 903-922.

MDG (1997), *Masonry Designers' Guide*, 2nd Edition, John H. Matthys, ed., The Masonry Society (Boulder, Colorado) and The American Concrete Institute (Farmington Hills, Michigan). (Chapter 14, "Connectors")

Melander, J. M., Ghosh, S. K., Dubovoy, V. S., Hedstrom, E. G. and Klingner, R. E. (1993), "Flexural Bond Strength of Concrete Masonry Prisms Using Masonry Cement Mortars," *Masonry: Design and Construction, Problems and Repair*, ASTM STP 1180, J. M. Melander and L. R. Lauersdorf, Eds., American Society for Testing and Materials, Philadelphia, PA.

Melander, J. M. and Ghosh, S. K., (1995), "Development of Specifications for Mortar Cement," *Proceedings* of the 1995 Symposium on Masonry: Esthetics, Engineering and Economy, Norfolk, Virginia, December 5, ASTM, West Conshohocken, Pennsylvania, pp. 88-99.

Melander, J. M. (1996), "Mortar Cement Gains Acceptance," *Magazine of Masonry Construction*, Aberdeen Group, Addison, Illinois, vol. 9, no. 8, pp. 371-375.

Meyer, C., Baxter, S. and Jin, W. (1996), "Potential of Waste Glass for Concrete Masonry Blocks," *Proceedings* of the Materials Engineering Conference Materials for the New Millennium, November 10-14, Washington, DC, American Society of Civil Engineers, New York, New York, pp. 666-673.

MSJC 1999: *Building Code Requirements for Masonry Structures (ACI 530-99 / ASCE 5-99 / TMS 402-99)*, American Concrete Institute, Farmington Hills, Michigan, American Society of Civil Engineers, Reston, Virginia, and The Masonry Society, Boulder, Colorado, 1999.

Nelson, R. L. and Barnett, R. E.(1997), "Autoclaved Aerated Concrete," *CMR Report*, Council for Masonry Research, Herndon, Virginia, vol. 9, no. 1.

Noland, J. L. (1990), "1990 Status Report: US Coordinated Program for Masonry Building Research, *Proceedings*, Fifth North American Masonry Conference, University of Illinois at Urbana-Champaign, June 3-6.

Priestley, M. J. N. and Seible, F. (1995), "Design of Seismic Retrofit Measures for Concrete and Masonry Structures," *Construction and Building Materials*, Butterworth-Heinemann Ltd., Oxford, England, vol. 9, no. 6, pp. 365-377.

Righetti, G. and Calliari, R. (1989), *Load-Bearing Masonry Structures: Calculations using Current Codes*, Consorzio Poroton Italia, Edizioni Lambda, Padova, Italy (in Italian).

RILEM 1993: *Autoclaved Aerated Concrete: Properties, Testing and Design*, RILEM Recommended Practice, RILEM Technical Committees 78-MCA and 51-ALC, E & FN Spon, London, 1993.

Schierhorn, C. (1994), "Integral-Water-Repellent Systems," *Magazine of Masonry Construction*, Aberdeen Group, Addison, Illinois, vol. 7, no. 11, pp. 514-516.

Schierhorn C. (1995), "Selecting Through-Wall Flashing," *Magazine of Masonry Construction*, Aberdeen Group, Addison, Illinois, vol. 8, no. 4, pp. 175-183.

Schultz, A. and Scolforo, M. (1991), "An Overview of Prestressed Masonry," *TMS Journal*, The Masonry Society, Boulder, Colorado, vol. 10, no. 1, pp. 6-21.

Shaw, G. (1996): "Innovative Development of Prestressed Masonry," *Proceedings*, ASCE Symposium in Conjunction with Structures Congress XIV April 15-18, Chicago, Illinois, American Society of Civil Engineers, Reston, Virginia.

Slaton, D. and Morden, M. R. (1996), "Issues in the Salvage and Reuse of Terra Cotta: Two Case Studies," ASTM Special Technical Publication Number 1258: *Proceedings* of the 1993 Symposium on Standards for Preservation and Rehabilitation, October 10-11, Dallas, Texas, ASTM, West Conshohocken, Pennsylvania, pp. 307-318.

Speare, R. S. (1995), "Recycling of Structural Materials," *Structural Engineer*, Institution of Structural Engineers, London, England, vol. 73, no. 13, pp. 220-222.

Tay, J.-H. and Show, K.-Y. (1994), "Municipal Wastewater Sludge as Cementitious and Blended Cement Materials," *Cement & Concrete Composites*, vol. 16, no. 1, pp. 39-48.

Wei, L., Naik, T. R. and Golden, D. M. (1994), "Construction Materials Made with Coal Combustion By-products," *Cement, Concrete and Aggregates*, ASTM, Philadelphia, Pennsylvania vol. 16, no. 1, pp. 36-42.

W. Lynn Beason[1] and A. William Lingnell[2]

Chapter 8

Emerging Uses for Window Glass

Abstract: Over the last several years the use of window glass has dramatically evolved. While window glass was once considered to be a secondary structural element that was largely ignored by the structural engineer, it has become a very important structural building component. This chapter presents general discussions relating to the different structural types and constructions of glass that are commonly encountered. In addition, discussions are presented relating to the use and specification of window glass for several different emerging design situations. While the information presented herein is not sufficient to accomplish a final structural window glass design, it is believed that sufficient information is presented to provide the designer with the proper direction to begin the structural design process.

Keywords: glass, window, wind, earthquake, blast, design

Introduction

Window glass has been in use for the past several centuries. Until recently, rules that govern the use of window glass largely have been based upon tradition and empirically generated information. This process worked reasonably well in the past because the uses and demands associated with window glass remained relatively constant. However, in the past few decades, the use of window glass has undergone a tremendous amount of change. Today window glass is used in sizes and applications that were unheard of a few decades ago. These changes have created confusion for designers who have not remained current in the area of window glass design. In many cases, designers lack basic information that is required to make informed decisions regarding the proper window glass product to be used for a particular application. Finally, when designers seek to collect relevant window glass design data they are faced with conflicting technical information that must be properly understood before the design process can be finalized.

The term "window glass design" encompasses a wide variety of factors including energy performance (Lingnell 1981), safety glazing concerns, aesthetic concerns, and basic structural adequacy questions. The information presented in this chapter is focused on basic structural adequacy questions. In particular, three emerging structural uses of window glass: windstorm resistant glass, earthquake resistant

---

[1] Associate Professor, Civil Engineering Department, Texas A&M University, College Station, Texas 77843-3136

[2] Principal, Lingnell Consulting Services, 1270 Shores Court, Rockwall, Texas 75087

glass, and blast resistant glass are examined.

Results of extensive damage documentation efforts have shown the effects of severe windstorms, earthquakes, and blasts on building performance. Through these efforts, window glass has been identified as a critical building element whose structural performance can be significantly improved. While there are controversies surrounding the proper course of action to be taken in a particular situation, it seems clear that certain modifications in the design and use of window glass can affect building occupant safety, and overall structural performance and utility of the building.

The purpose of this chapter is to present the reader with critical background information concerning the availability and properties of currently available types and constructions of window glass and to identify controversial topics relating to window glass design. While the level of information presented is not sufficient to develop final window glass designs, it is believed that the information will allow a reasonably informed designer to select the most appropriate window glass for a particular application, further the evaluation process, and make initial plans for its design and use.

The remainder of this chapter is divided into four major sections. The next section presents a review of the two basic types of window glass that are in widespread use today. These types of glass include annealed float glass and heat-treated glass. This is followed by a section that presents a review of the three major types of glass constructions that are in common use including insulating glass, laminated glass, and filmed glass. This is followed by a section that discusses various design and analysis elements that must be considered in developing structural designs for window glass subjected to windstorm, earthquake, and blast loadings. The final section presents general conclusions and observations relating to the structural design of window glass.

Types of Window Glass

The primary flat glass product produced in the United States today is annealed float glass or simply annealed glass. Annealed float glass is a flat glass product that has a minimum level of residual stress. For certain applications, annealed float glass can be subjected to a heat-treatment process that locks in a high level of residual surface and edge compression. This high level of residual surface and edge compression greatly increases the strength of heat-treated glass when it is exposed to certain types of loadings. The properties and uses of annealed and heat-treated glass are discussed below. (GANA 1997)

Annealed Float Glass

Annealed float glass or simply annealed glass is the most widely used type of glass. Annealed float glass is manufactured in a process where molten glass is poured continuously onto a bed of molten tin. The molten glass tends to seek a uniform thickness as it floats on the surface of the molten tin. Because the melting point of the tin is less than that for glass, the glass slowly cools and solidifies as it floats across the bed of molten tin. After the glass exits the tin bath, it moves into an annealing lehr where it continues to be slowly cooled so that residual stresses are minimized in the glass. This process results in a product that is very flat with nearly parallel surfaces. The annealed float glass process has replaced virtually all other types of flat glass processes in commercial applications including the plate glass and the sheet glass processes. (GANA 1997)

Annealed glass is a brittle elastic material whose strength is controlled by the interaction of tensile stresses with stress raising flaws (Beason 1980, Beason and Morgan 1984, and Beason, et al 1998). Stress raising flaws are located within the body of the glass and on the glass surface, but it is usually the surface flaws that are of most concern. While the term flaw carries with it a certain negative connotation, the flaws under discussion are an inevitable result of the manufacturing and handling processes. In addition, once a window is placed into service, it is exposed to a wide variety of mechanical exposures that create additional flaws. These in-service flaws tend to be less frequent in occurrence, but more severe than the flaws that develop during manufacturing (Beason, et al 1998).

Annealed glass is at its highest strength as it exits the manufacturing process. As the glass begins the fabrication, handling, and distribution processes, it is subjected to an increasing number of mechanical exposures that introduce new and more severe flaws on the surfaces of the glass. The strength of annealed glass quickly converges to a level that is representative of the glass handling and fabrication processes and the in-service environment (Beason, et al 1998).

Results of uniform load tests of different sets of in-service glass plates show that the strength of annealed glass that has been in service for a period of ten to twenty years, is much less than the strength of freshly manufactured glass of the same geometry and type (Beason 1980, Abiassi 1981, Norville and Minor 1985). These results show that in some cases the strength of in-service annealed glass can be as much as 50 percent less than the strength of freshly manufactured annealed glass. The fundamental question that is introduced by these observations is what causes the strength reductions.

In unpublished research conducted at Texas A&M University, it has been found that there are no major differences in the strength of freshly manufactured glass specimens and in-service glass specimens when both types of specimens are subjected to the same artificially induced surface flaws, the same environmental exposures, and the same types of loads. In this effort, 203 beam specimens were prepared from glass taken from a 60-year old building. These glass beams were then artificially abraded, subjected to a specified set of environmental conditions, and tested to failure by the application of a four point bending load. The severity of the artificial abrasions was such that all failures initiated in the abrasion zone away from the edges of the beam specimens. Then 205 beam specimens were prepared from freshly manufactured glass of the same nominal thickness. These specimens were subjected to the same treatment and testing afforded the in-service glass. It was found that the mean failure stress of the 205 freshly manufactured glass beams was about 63 MPa while the mean failure stress of the 203 in-service glass beams was about 60 MPa. It was concluded that while this difference is significant from a statistical standpoint, a difference of 5% or less is probably not significant from an engineering standpoint. Therefore, it was concluded that the reduction of glass strength with in-service exposure is not the result of chemical or material changes that occur over time. The most likely alternative explanation is that the cumulative effects of mechanical exposures are responsible for the primary reduction of glass strength with in-service exposure. (Schrader 1982)

The flaws that seem to be most important in the investigation of the strength reduction of in-service annealed glass are surface flaws. This is the case because surface flaws tend to be more severe than internal flaws. In addition, the magnitudes of the tensile stresses induced by typical loadings are greater on the surfaces of glass plates than in the plate's interior. Hence, almost all in-service glass failures initiate on the surface or edges of the glass in an area of relatively high tensile stress. Therefore, the strength of in-service glass tends to be dominated by the character and distribution of in-service surface flaws.

It is well recognized that the surface of glass is chemically affected by exposure to water vapor at levels that are common in the atmosphere. The rate of attack of water vapor on a glass surface is very slow so that its effects are usually apparent only on a microscopic level with no obvious macroscopic changes. However, it is known that exposure to common levels of water vapor can significantly alter the strength of glass. If glass is exposed to water vapor in the presence of tensile stress, there will be an overall reduction in glass strength. This phenomenon is usually termed static fatigue. (Mencik 1992).

Figure 1 presents the variation of normalized failure stress with load duration based on conventional understandings of the static fatigue performance of annealed glass (Beason and Morgan 1984). The information presented in Figure 1 is normalized to a 60-second duration constant loading. Therefore, if the horizontal axis of the graph presented in Figure 1 is entered for a 60-second duration, it is found that the normalized failure stress is 1.0. If the horizontal axis is entered with a 1-hour (3600 seconds) duration, the normalized failure stress is about 0.77. This means that the 1-hour duration mean failure stress is about 23% less than the 60-second duration mean failure stress for a particular situation. Thus, if tests have shown that the mean 60-second duration failure stress associated with a particular group of specimens is 40 MPa, then the estimated mean 1-hour duration failure stress would be about 31 MPa for similar specimens. If the duration of the loading is less than 60-seconds, the normalized failure stress increases. Historically, glass thickness selection charts have been based on 60-second load durations, although there is currently some pressure to reduce the design duration to 3 seconds. The normalized failure stress increases by about 20% as the duration reduces from 60 to 3 seconds.

The phenomenon of static fatigue has been erroneously credited for the long term in-service strength reduction discussed previously. However, it must be noted that the phenomenon of static fatigue occurs only if the surface flaws are simultaneously subjected to the effects of water vapor and tensile stress. If either component is absent, there is no static fatigue. In fact, there is significant evidence available in the literature to suggest that if surface flaws are exposed only to water vapor in the absence of tensile stress, the strength of the glass may increase (Mencik 1992). There is no evidence to suggest that the strength of glass is sensitive to cyclic fatigue in the classical sense.

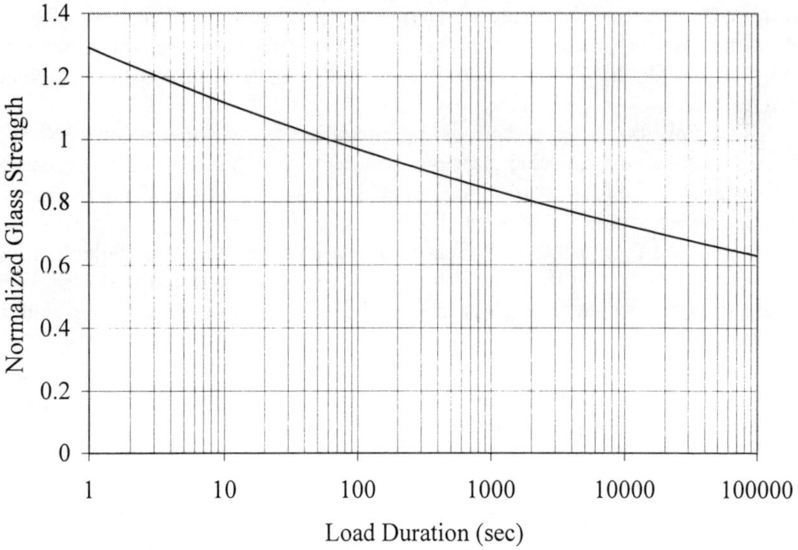

Figure 1 - *Variation of Glass Strength Normalized to 60-Second Duration Loading*

The uniform pressure resistance of annealed glass is often the primary structural criterion used in the selection of the proper thickness of glass to be used in a particular situation. To model the uniform pressure resistance of annealed glass it is necessary to be able to model the interaction between the surface tensile stresses and the in-service surface flaws as a function of load duration. One accepted procedure to accomplish this is through the use of a glass failure prediction model (GFPM) developed by Beason (Beason 1980, Beason and Morgan 1984, Beason, et al 1998). The GFPM is a theoretical formulation that allows the probability of failure of a laterally loaded annealed glass plate to be expressed in terms of the characteristics of the surface flaws and the distribution of surface tensile stresses. The GFPM incorporates the effects of the various factors discussed above such as load duration, plate geometry, and glass surface condition. In addition, the GFPM incorporates the results of a geometrically nonlinear plate analysis to calculate the variation of surface tensile stresses with the magnitude of the lateral load.

Annealed glass thickness selection charts based on the GFPM are presented for general design purposes in ASTM E 1300 (ASTM 1998). Figure 2 presents a typical glass thickness selection chart for 6 mm glass. The chart is entered with the length and width of the glass plate and the allowable design load is read from the chart. The resulting glass thickness recommendation is based on a 60 second duration loading. ASTM E 1300 contains similar charts for all common annealed glass thicknesses.

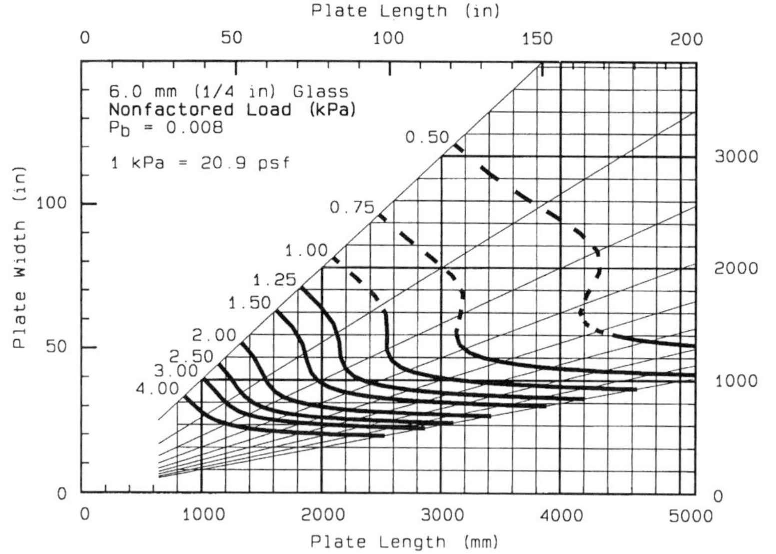

Figure 2 - *ASTM E1300 Glass Thickness Selection Chart for 6 mm Glass*

While the annealed glass thickness charts presented in ASTM E 1300 are becoming the most commonly used glass thickness selection charts, they are not the only ones available. Some building codes and some manufacturers continue to present the traditional annealed glass thickness selection chart. The traditional glass thickness selection chart is based on empirical information generated in the late 1950's and early 1960's (Hershey and Higgins 1973). Figure 3 presents an example of the traditional annealed glass thickness selection chart. This chart is entered with the area of the window glass and the magnitude of the uniform pressure. Then, the required minimum thickness of glass is selected. Again the resulting glass thickness recommendation is based on a 60-second duration loading. (Beason 1980)

In contrast to the glass thickness selection charts presented in ASTM E 1300, the traditional glass thickness selection chart is based on an empirical representation of strength data collected in the testing cf freshly manufactured annealed glass specimens (Hershey and Higgins 1973). While the empirical data set used to develop the traditional glass thickness selection chart included glass plate aspect ratios (length of rectangular glass divided by width) ranging from 1:1 to 1:5, the traditional chart is presented in such a way that it appears that the strength of glass depends only on the area and thickness of glass and not aspect ratio. The E 1300 glass thickness selection charts are more refined with respect to the influence of aspect ratio. The minimum glass thickness determined using the traditional glass thickness selection chart is frequently less than the corresponding glass thickness determined using the glass thickness selection charts presented in ASTM E1300 (Beason and Norville 1989). Therefore, the designer must be careful to assure that window glass designs to resist specified lateral pressures are conservative.

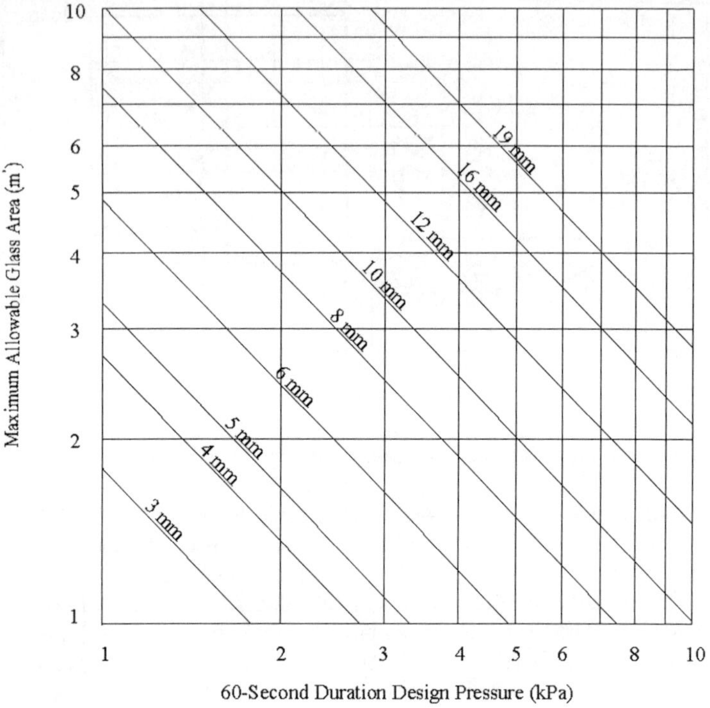

Figure 3 - *Traditional Glass Thickness Selection Chart*

The other major structural design consideration for annealed glass is its response to thermal stresses. Thermal stress fractures are the result of uneven temperature distributions across the surface of the glass plate. The most common cause of uneven temperature distribution is unequal heating of the glass as the result of exposure to sunlight. Thermal stress fractures tend to originate at the edges of glass plates as the result of the interaction of thermally induced tensile stresses with edge flaws or edge damage. In addition, severe surface damage or flaws can also trigger thermal stress fractures. This problem is exacerbated by the heat absorption characteristics of the glass, the magnitude of the edge bite, improperly installed interior shading devices, exterior shading conditions, as well as other factors. Currently there is no unified failure theory that can be used to design for thermal exposure conditions. Rather, the designer must rely on empirical information presented by glass manufacturers in evaluating the risks associated with thermal exposures.

Finally, one of the biggest problems associated with the use of annealed glass is that when it does break, it tends to break into a small number of relatively large pieces of glass. These large shards of glass have very shape edges that present a life-threatening hazard to building occupants or nearby pedestrians. Therefore, monolithic annealed glass is never considered appropriate for safety glazing applications.

Heat-Treated Glass

The strength of annealed glass can be greatly increased by subjecting it to a heat-treatment process that locks in a residual stress distribution into the glass. In this process, annealed glass is heated to a temperature that is near its softening point and then quickly quenched. In the quenching process, the surface of the glass is cooled quicker than the interior of the glass. This differential cooling causes the surface of the glass to become rigid while the interior remains soft. When the interior of the glass eventually cools, a residual stress pattern is locked into the glass. (GANA 1997)

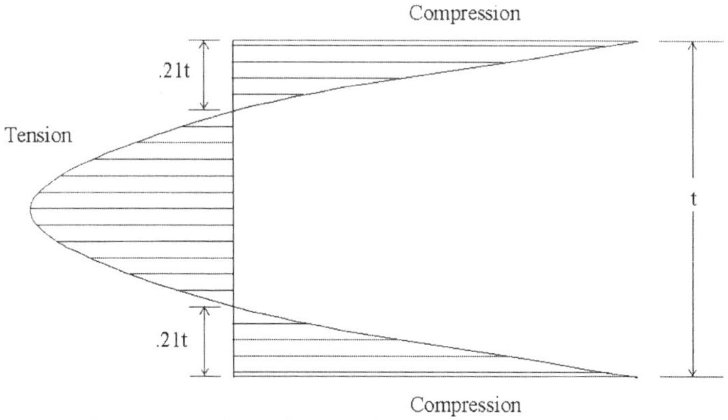

Figure 4 - *Idealization of Residual Surface Compression in Heat-Treated Glass*

Figure 4 presents an idealization of the variation of residual stress through the thickness of heat-treated glass. As shown in Figure 4, the exterior surfaces of the glass are subjected to compressive stresses, while the interior of the glass is subjected to tensile stresses. Because the cross-section of the glass must remain in equilibrium, it can be shown that the total resultant compressive force acting through the thickness of the heat-treated glass must be equal in magnitude to the total resultant tensile force. It is usually assumed that the variation of the residual stress distribution through the thickness of the glass is parabolic. If this assumption is coupled with the necessary equilibrium and boundary conditions, it can be shown that the magnitude of the maximum residual interior tensile stress is equal to half of the magnitude of the maximum residual surface compression. This heat-treatment process does not affect the elastic properties of glass. (Mencik 1992, Lingnell 1994a)

The net surface tensile stress is defined as the difference between the applied tensile stress and the residual surface compression. Therefore, before a flaw on the surface of heat-treated glass is exposed to a net surface tensile stress, the applied mechanical tensile stress must exceed the residual surface compression. Hence, the strength of heat-treated glass is significantly greater than the strength of annealed glass. It is probably reasonable to assume that heat-treated glass is not sensitive to the effects of static fatigue until the net surface stress becomes tensile. However, additional research needs to be conducted to fully understand the static fatigue behavior of heat-treated glass. In the interim, it is recommended that the annealed glass static fatigue relationship be applied to the net tensile stress experienced by heat-treated glass.

The amount of the strength increase experienced by heat-treated glass is dependent on the magnitude of the residual surface compression. The two most common classifications of heat-treated glass are referred to as heat-strengthened and fully tempered glass. By definition, heat-strengthened glass must have a residual surface compression greater than 24 MPa but less than 52 MPa and fully tempered glass must have a minimum level of residual surface compression greater than 69 MPa (ASTM 1997a). Common rules of thumb that are often used to conservatively quantify the uniform pressure resistance of heat-treated glass are that heat-strengthened glass is 2 times as strong as annealed glass of the same geometry, and fully tempered glass is 4 times as strong as an annealed glass of the same geometry. In specific instances the relative strength increase can be significantly greater.

If it is necessary, to achieve a more accurate assessment of the strength of heat-treated glass, it is possible to use a modified version of the GFPM that is based on the net surface tensile stress experienced by a glass plate. The validity of the modified GFPM has been demonstrated though comparisons of strength predictions for the strength of heat-treated glass with experimentally measured strengths. (Beason 1993)

The breakage characteristics of heat-strengthened glass are often similar to the breakage characteristics of annealed glass. Hence, heat-strengthened glass is not recommended for use in safety glazing applications. However, when fully tempered glass breaks, it usually breaks into a large number of relatively small pieces of glass that are relatively innocuous when compared with the large sharp glass shards associated with the breakage of annealed or heat-strengthened glass. In addition, it is well documented that fully tempered glass is more resistant to impacts than annealed glass. Hence, fully tempered glass is frequently used in safety glazing applications.

One problem associated with the use of heat-treated glass is frequently referred to as spontaneous breakage. Spontaneous breakage of heat-treated glass can be defined as the catastrophic failure in the apparent absence of an extreme loading situation. Spontaneous breakage is usually associated with heat-treated glass with high levels of residual surface compression. Hence, fully tempered glass is generally more susceptible to spontaneous breakage than heat-strengthened glass. Spontaneous breakage is believed to be the result of the introduction or extension of an unstable flaw into the interior tension zone of heat-treated glass. When this happens, fractures propagate through the glass driven by the residual stress distribution. The higher the magnitude of the residual stresses, the more extensive is the fracture pattern.

Based upon information presented by the Glass Association of North America, the unstable flaw that triggers a spontaneous failure can be the result of surface or edge damage or it can be the result of internal impurities in the glass. Critical edge or surface damage can be caused by deep scratches, gouges, severe weld spatter, impacts, glass to metal contact, or other sorts of mechanical abuse. While the edge or surface damage may not initially be severe enough to initiate failure, the flaws introduced can become increasingly more severe if the glass is subjected to cyclic wind or thermal loadings. Spontaneous breakage occurs when the edge or surface damage penetrates the compression layer of the heat-treated glass. (GANA 97)

There are many different types and sizes of internal impurities that can occur in glass. There are roughly 50 different types of inclusions or impurities that have been identified in the manufacture of glass (GANA 97). The vast majority of these impurities do not affect the structural performance of heat-treated glass. However, it is known that a few relatively rare impurities or inclusions can cause an unstable flaw growth in the tension zone of heat-treated glass thus leading to a spontaneous failure.

Conventional wisdom in the glass industry suggests that nickel sulfide is one very important impurity that can lead to spontaneous breakage. It is widely believed that extreme temperature variations in the glass can result in phase changes in a nickel sulfide impurity that will cause an increase in the volume of the impurity. This volume increase creates an internal tensile stress that is additive with the internal residual tensile stress at the point of the impurity. If the net internal tensile stress reaches a critical level, it can trigger a spontaneous failure that is seemingly unrelated to external loadings. The larger the internal tensile stress, the greater the risk associated with spontaneous breakage. Hence, fully tempered glass has a much greater probability of spontaneous breakage than heat-strengthened glass. Although, spontaneous breakage has been observed by the writers for both heat-strengthened and fully tempered glass, the probability of spontaneous breakage is very small in heat-strengthened glass with a residual surface compression less than 52 Mpa. Therefore, where fully tempered glass is not required by building codes and/or design loads, heat-strengthened glass is the preferred product.

While there is no widely accepted procedure that can be used to identify potentially damaging internal flaws, it is possible to reduce the probability associated with spontaneous breakage. This is done by subjecting the heat-treated glass to a heat-soak procedure. In this procedure, heat-treated glass plates are subjected to an elevated temperature in the general range of 200-260° C or greater for a specified period (usually a few hours). The conditions associated with the heat-soak procedure are intended to be more severe than normal service conditions. Thus, if a heat-treated glass plate survives the heat-soak process it has a significantly reduced potential for spontaneous breakage. The precise variation of temperature used in the heat-soak process varies from company to company and is generally considered to be proprietary. The heat-soak procedure can be used to test all heat-treated glass plates for a particular job, or a fabricator can heat soak randomly selected samples of glass for general quality control.

Glass Constructions

In certain applications, it makes sense to use monolithic glass to fabricate a variety of different glass constructions for various architectural applications. For instance, multiple pieces of monolithic glass can be fabricated into laminated glass or insulating glass. Each of these products has advantages over the use of monolithic glass and will be discussed. In certain applications, the post breakage performance of monolithic glass plates can be enhanced through the use of polyester (PET) films that are typically applied after the glass has been installed. While there are a number of miscellaneous types of glass that may also be encountered in the marketplace including wired, textured, and patterned glass, this discussion is limited to the three most common glass constructions including insulating, laminated, and filmed glass. The remainder of this section is devoted to a discussion of these three different glass constructions.

Insulating Glass

Most insulating glass (IG) units are manufactured by enclosing a sealed gas space between two pieces of glass that are separated by a perimeter spacer as shown in Figure 5. The two glass plates are typically referred to as the inboard and outboard lites. The outboard lite is exposed to the exterior building environment while the inboard lite is exposed to the interior building environment. The thickness of the gas space is controlled by the thickness of the spacer that is inserted between the glass plates along the perimeter edges. Elastomeric sealants are applied around the perimeter of the IG unit during the fabrication process so that the glass plates and spacers are bonded together and the gas space is sealed. A desiccant is incorporated into the IG system, usually inside the perimeter spacer, to adsorb water vapor that migrates into the air space through the seal during the life of the IG unit. (Beason 1986a, Beason 1986b)

The purpose of the two glass plates in the IG construction is to reduce heat flow through the window by providing two vapor barriers instead of one. In addition, it is possible to further reduce heat flow through the window by replacing the air in the gas space with an inert gas such as argon. In addition, there are a wide variety of low-emissivity or solar-control coatings that can be applied to the surfaces of the IG lites to further control heat flow. A similar technology can be used to fabricate IG units with more than two glass plates, although the use of triple plate IG units is usually restricted to areas that experience extreme cold conditions in the northern latitudes. The remainder of this section is devoted to a discussion of the structural performance of IG units fabricated with two glass plates and one gas space.

When an IG unit is sealed, the gas space pressure is equal to the atmospheric pressure at the time that it was sealed so that both the inboard and outboard lites are exposed to no load. If the IG unit is subsequently exposed to a lateral load, the outboard lite will deflect into the gas space, causing a gas space volume reduction that increases the gas space pressure. The increased gas space pressure causes the inboard lite to deflect. As the inboard lite deflects, the gas space volume is increased and the pressure within the gas space is reduced. When the IG unit reaches equilibrium, the lateral load is shared between the inboard and outboard lites. The distribution of the load between the IG lites depends upon the length, width and thickness of the glass plates and the gas space thickness.

An IG load sharing analysis has been developed and experimentally verified to model the load sharing characteristics of IG units (Beason 1986b). Use of this analysis has shown that if the two IG plates have equal thicknesses and are sized to resist common architectural loads, the plate through which the load is applied will typically resist slightly more load than the other plate. A common situation might be that the loaded glass plate would resist 51% of the applied load while the other plate would resist 49% of the applied load.

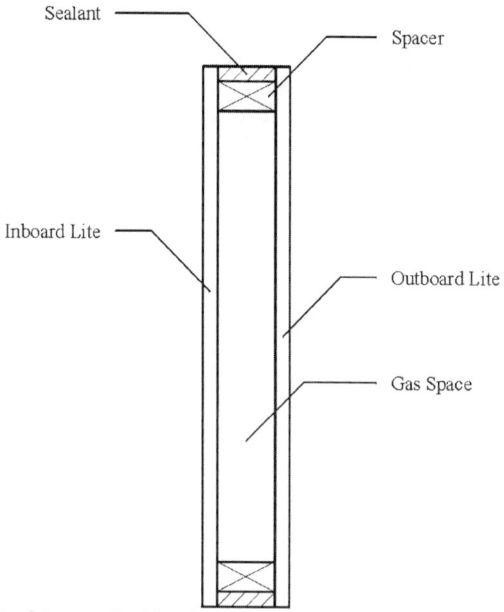

Figure 5 - *Idealization of Insulating Glass Unit*

If an IG unit is exposed to both a lateral load and variations of atmospheric pressure and gas space temperature changes, the load sharing characteristics of the IG unit can be significantly altered. For instance an increase in the gas space temperature or a decrease in the atmospheric pressure will change the net loads acting on the IG lites so that the IG unit tends to bulge. The bulging effect pre-loads the IG lites so that if a lateral load is subsequently applied to the IG unit, the net load resisted by the outboard lite will be reduced while the net load resisted by the inboard lite will be increased. If an IG unit is exposed to an increase in atmospheric pressure or a decrease in air space temperature, the effect on the IG unit will be opposite. The temperature and pressure effects are more pronounced for small IG units with relatively stiff plates. The IG load sharing analysis discussed above is capable of modeling the effects of the temperature and pressure variations.

A failure prediction model for IG units has been developed to model the strength of IG units (Beason 1986b). This failure prediction model incorporates the GFPM (Beason and Morgan 1984) and the results of the IG load sharing analysis discussed above (Beason 1986a). Results generated with the IG failure prediction model show that if a typical IG unit is fabricated with two glass plates of the same type and thickness it is reasonable to assume that the total load capacity of the IG is unit is equal to about 1.8 times the load carrying capacity of either of the glass plates (Beason 1986b). This factor is incorporated in ASTM E 1300 (ASTM 1998).

Laminated Glass

Laminated glass is manufactured by bonding multiple layers of glass together through the use of adhesive interlayers. The glass that is used to fabricate laminated glass can be either annealed or heat-treated depending upon the intended use of the finished product. The adhesive interlayer used to bond the layers of glass together can be anyone of several different types; however, a mixture of plasticizers and polyvinyl butyral (PVB) is by far the most common interlayer material in use for the laminated glass interlayer. The most common laminated glass configuration involves two pieces of glass of the same thickness and type that are bonded together with a PVB-based interlayer. During the fabrication process, the PVB-based interlayer is placed between the two glass plates. Then, the entire assembly is subjected to a combination of heat and pressure so that a bond is formed between the interlayer and the glass. Figure 6 presents a cross-section of a typical laminated glass unit. (GANA 1997)

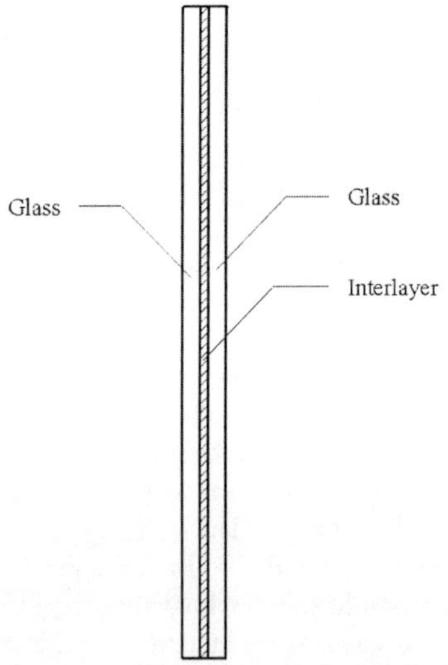

Figure 6 - *Idealization of Laminated Glass*

Laminated glass has several advantages over monolithic glass. One advantage is that the introduction of the interlayer provides for improved acoustic insulation for buildings. In addition, solar transmission, absorption, and tinting of a laminated glass unit can be altered through the use of appropriate interlayer additives. However, from a structural standpoint, the most important advantage is that if a laminated glass unit is broken, the interlayer tends to adhere to the glass fragments reducing the tendency for the glass fragments to exit the window framing system. This can lead to a more desirable post-breakage result than occurs with monolithic glass. This is perhaps the most advantageous characteristic of laminated glass.

The structural properties of laminated glass are a function of the individual properties of the glass and the PVB-based interlayer. While the elastic properties of glass remain relatively stable over normal temperature ranges, the elastic properties of the PVB-based interlayer vary significantly with the percentage of plasticizer and the ambient temperature. This variation of the elastic properties of the PVB interlayer with plasticizer ratio and temperature causes the structural behavior of laminated glass to vary significantly with temperature. (Beason et al 1998)

Both the shear modulus, G, and the modulus of elasticity, E, reduce as the temperature of the PVB-based interlayer increases. As with most amorphous polymers, the rate of change of these elastic properties with respect to temperature undergoes a radical change at a point that is traditionally referred to as the glass transition temperature, T_g, or simply the transition temperature. The transition temperature serves as an important performance boundary for the interlayer. The PVB-based interlayer tends to be stiff, leathery, and exhibit a limited amount of ductility at temperatures that are well below the transition temperature; while, the PVB-based interlayer tends to be soft, ductile, and rubbery at temperatures that are above the transition temperature. At temperatures that are at or just below the transition temperature the PVB-based interlayer tends to be very ductile with a performance that exhibits a very viscous response. (Beason et al 1998)

The transition temperature of pure PVB is reported to be on the order of 70° C. However, as the percentage of plasticizer that is incorporated in the PVB-based interlayer increases, the transition temperature decreases to value that is limited by the transition temperature of the plasticizer. While there is very little mention of the glass transition temperature of the PVB-based interlayer in the open literature, it has been reported that one PVB manufacturer adjusts the level of plasticizer so that the transition temperature of the interlayer is about 25° C. Informal discussions with interlayer manufacturers suggest that this value of the transition temperature is reasonably typical of most commonly available architectural grade PVB-based interlayers. A 25° C transition temperature is at the upper end of room temperature, so that when impact tests such as those associated with safety glazing certifications are conducted at room temperatures, the laminated glass will absorb an optimum amount of impact energy. However, some of the desirable impact performance of laminated glass will be significantly altered at temperatures that are below the 25° C transition temperature. (Fariss 1993)

At temperatures above its transition temperature, the modulus of elasticity of the PVB-based interlayer is as much as 5 orders of magnitude less than that of glass. At temperatures near 0° C, the elastic modulus of the PVB-based interlayer continues to be almost 3 orders of magnitude less than that of glass. Figure 7 presents the variation of the shear modulus of the PVB-based interlayer as a function of temperature as determined experimentally by Hooper (1973). Examination of Figure 7 shows that the shear modulus of the interlayer reduces by a factor of almost 100 as the temperature increases from 0° to 25° C. Then the shear modulus reduces only slightly as the temperature increases from 25° to 50° C. The change in slope of the shear modulus vs. temperature curve at about 25° C indicates that the transition temperature of architectural PVB-based interlayer is about 25° C as reported by Fariss (1993).

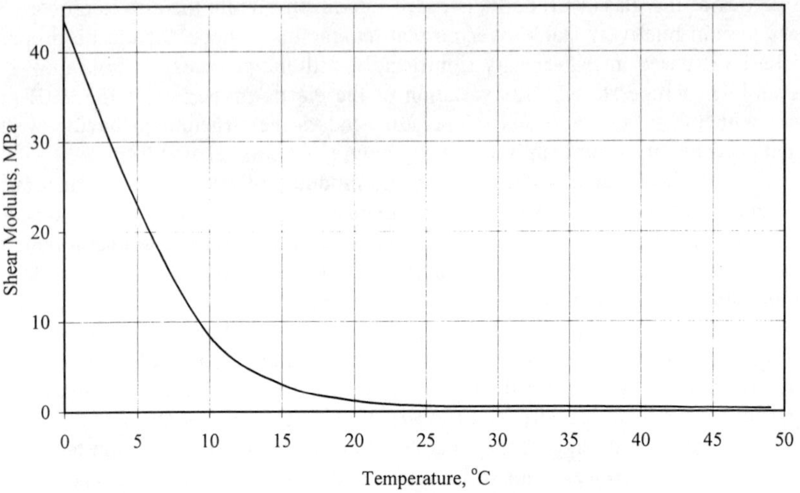

Figure 7 - *Variation of Shear Moudulus of PVB-Based Interlayer*

With differences this large between the elastic properties of glass and the PVB-based interlayer, use of conventional composite construction theory is precluded for use with laminated glass. To properly model the behavior of laminated glass it is necessary to employ analysis techniques that are not dependent on the assumption that plane sections through the thickness of laminated glass before loading remain plane after loading. Therefore, care must be exercised in selecting an appropriate analysis technique for laminated glass because many composite structural analysis procedures in common use are based on the "plane sections remain plane" premise. (Beason et al 1998)

Analysis techniques that accommodate the behavior of laminated glass have been developed for both laminated glass plates and beams. However, these solution techniques tend to be highly specialized and difficult to apply. In most cases, it has been found by researchers at Texas A&M University that properly formulated finite element models using available software packages provide good solutions for laminated glass applications. It is recommended that finite element models be used to determine the structural response of laminated glass for design purposes. (Beason et al 1998)

While it is clear that finite element models provide the best tool for modeling the structural response of laminated glass, in some cases it is not possible to allocate the resources required to properly analyze the laminated glass unit. Therefore, simplified approximations are often used for general design purposes. If the PVB-based interlayer is above its transition temperature, 25° C, then it is very soft and flexible so that the shear transfer between the two glass plates that make up the laminated glass is minimal. In such cases, the performance of laminated glass tends toward the performance of a layered system where the laminated glass plates are in contact with each other but there is no shear transfer. In this case, the uniform pressure resistance or stiffness of the laminated glass unit is conservatively approximated by summing the uniform pressure resistances or stiffnesses of the individual plates.

If the PVB-based interlayer is at a temperature that is near or below 0° C then the stiffness of the PVB-based interlayer is sufficient to assure substantial shear transfer between the two glass plates that make up the laminated glass unit. In this case, the performance of the laminated glass unit approaches that of a monolithic glass plate of the same overall thickness as the laminated glass unit. These layered and monolithic laminated glass models provide performance limits for most common types of laminated glass.

In most cases, use of the layered model will result in lower load capacities than the use of the monolithic model. However, there are rare instances where it has been reported that the monolithic model results in a lower estimated load capacity than the layered model (Behr, et al 1993). This coupled with the fact that most windows are subjected to a wide range of operating temperatures during their in-service life leads to the conclusion that each design should be checked using both the layered and the monolithic models before the design is finalized. This is particularly important in the design of insulating glass units that employ laminated glass for one of the IG lites. In this case, the stiffer the laminated glass unit, the more load that it is required to resist. Thus, it depends on the overall geometry of the unit as to whether the layered model or the monolithic leads to the most conservative design. If it is important that more exact solutions be developed for a particular application it is recommended that an appropriate finite element model be used as recommended earlier.

It should be noted that results of research conducted by researchers under the sponsorship of the PVB-based interlayer manufacturers suggest that the performance change between the monolithic and layered models doesn't occur until 50° C instead of 25° C (Behr et al 1993). If this were the case, it would be possible to treat laminated glass as though it were monolithic glass for many common design situations. This would mean that it would be possible to retrofit monolithic glass with laminated glass of the same thickness without having to modify the window frames to accept a thicker product. However, independent research has not confirmed the interlayer manufacturer's results (Beason et al 1998). Designers must be sensitive to this conflict.

As stated previously in this section, one of the major benefits associated with the use of laminated glass is an improved post-breakage performance. While there are various sets of favorable test results available, there are no accepted techniques that allow the improvement in the post-breakage performance to be quantified. Therefore, if it is necessary to quantify the post-breakage performance of laminated glass for a particular application, it continues to be necessary to perform full-scale tests that accurately model the intended application. Factors that are important to the post-breakage performance of laminated glass include the thicknesses of the glass, the thickness of the interlayer, the stiffness of the supporting frame, and the manner in which the laminated glass is attached to the window frame. This testing must be accomplished on a case by case basis.

Filmed Glass

Window film consists of a thin polyester (PET) film that is coated with a suitable adhesive that allows the film to be securely fastened to the surface of glass. PET film can be applied to the glass during fabrication of the window or it can be retrofitted on existing windows. Usually the film is applied to an interior surface of the glass so that the film is not exposed to the exterior building environment. Window film can be fabricated with a wide variety of different characteristics that affect the tinting, solar reflection, and absorption of the glass unit. Depending upon the intended use of the window, the film can be fabricated out of a single layer of film or multiple layers of film can be laminated together to form a thicker product. (AIMCAL 1996)

As described above, the window film is adhered directly to the surface of the glass. Typically, the thickness of the window film is only a few percent of the glass thickness, often less than one percent. Further, the modulus of elasticity of the PET window film is much less than the modulus of elasticity of window glass. Therefore, both the stiffness and strength of filmed window glass is not substantially different to the stiffness and strength of unfilmed glass of the same thickness. Thus, the only structural benefit that is to be derived from the use of filmed glass is an improved post-breakage performance of glass in much the same way that the PVB-based interlayer improves the post-breakage performance of laminated glass. (AIMCAL 1996)

The degree to which the post-breakage performance is enhanced for filmed window glass depends upon many factors including the thickness and construction of the window film. The thinner single layer films are typically used to alter the appearance and light transmission characteristics of glass, while thicker multilayer films are usually employed to provide an enhanced post-breakage performance. The film can be trimmed so that its edge is flush with the edge of the supporting window frame. This type of application is typically referred to as a daylight application. In addition, the film can be applied so that the window film is continued past the edge of the glass and mechanically attached to the window frame. This type of application is often referred to as an anchored film application.

There is evidence to suggest that even daylight applications of solar control film provide an enhanced post breakage performance for window glass. In this case, the glass fragments may be contained and controlled after breakage depending upon the post breakage loading situation. However, if it is desired to develop a repeatable level of post breakage performance enhancement, it is generally deemed necessary to employ an anchored film application. At the present, there are no widely accepted design procedures for quantifying the improvement of the post breakage performance short of employing compliance testing. Thus testing has to be conducted to determine the performance of each type of window that is fabricated with a specific combination of materials and subjected to a specific loading. As is the case with laminated glass, filmed glass should not be used in a high threat area until adequate performance testing has been completed.

Emerging Uses of Glass

Over the past several years, the expectations associated with the use of glass have changed significantly. Prior to the 1980's it was widely accepted that glass failure is a natural consequence of extreme loading situations. As a result, the consequences of glass failure were tolerated by society. However, in the last several years, it has become obvious that the performance of window glass under extreme loadings can drastically affect occupant safety and the overall structural performance of a building. As a result, new design criteria and procedures are under consideration to minimize the negative effects associated with extreme loadings. The two most common types of natural disasters that affect the performance of window glass are earthquakes and extreme wind loadings. In addition, it is quickly becoming accepted that there is an increasing threat that a building will be subjected to the effects of blast loadings. Blast loadings can result from either accidental situations or terrorist actions. In each of these cases the performance of the window glass can greatly affect the safety of building occupants, the overall level of damage to the building, and the degree to which the building can provide uninterrupted service. Each of these topics is discussed below.

Windstorm Applications

The term windstorm encompasses a wide variety of different but related events. Included in this list are severe straight-line windstorms, tornadoes, downbursts, thunderstorm outflows, and hurricanes. In all of these cases, windows are subjected to the effects of uniform lateral pressure as the result of moving air impinging on the surface of the glass. Depending upon the circumstances and the speed of the wind, the structure may be subjected to a barrage of windborne debris. Considerable effort has been expended in the past to develop design procedures and test methods to assure that window glass can be designed to withstand the effects of wind pressure. The culmination of this effort is presented in ASTM E1300 (ASTM 1998). However, the effects of windborne debris have largely been ignored in formal design procedures until the early 1990's.

Observations that have been presented by researchers since the early 1970's show that as sustained wind speeds exceed a threshold level of perhaps 40 m/s, the winds begin to actively transport a wide variety of projectiles. Studies have shown that these windborne missiles can be quickly accelerated by the force of the wind to speeds approaching half the sustained speed of the wind or more. If missiles impact window glass at speeds of this magnitude, a missile induced failure can result. In fact, it is well documented that the majority of glass failures in windstorms are the result of missile impact and not excessive lateral loading caused by wind pressures. For purposes of structural evaluation, windborne missiles can be divided into two different groups: large missiles and small missiles. (Minor and Beason 1976, Minor et al 1978, Beason et al 1984)

The small missile group includes small objects and debris with masses of only a few grams. Many damaging small missiles are found in large quantities on the roofs of urban structures. Small missiles located on urban roofs are particularly dangerous because their initial elevation provides for easy injection into a wind stream. The failure mechanism associated with small missile impact tends to be a localized stress wave phenomenon. Hence, glass failures result at energy levels that would normally not be expected to cause structural failure of glass. For a small missile to cause a critical stress wave it is necessary for it to be hard with respect to glass. Objects such as brick bats, stones, metallic objects, etc. are all capable of inducing localized stress waves in the glass upon impact. Roof gravel has been identified as one of the most common sources of damaging small missiles that is commonly available in urban areas. Loose roof gravel is available in large quantities on both low-rise and high-rise buildings. Results of small missile impact tests show that all types of monolithic glass including annealed, heat-strengthened and fully tempered glass are subject to failure as the result of small missile impact. (Minor and Beason 1976 and Minor et al 1978)

Large missiles are much heavier objects that are capable of causing glass failure on the basis of excessive impact energy. Large missiles include loose debris at the ground level or on roofs such as trashcans, lawn furniture, pieces of sheet metal, etc (Minor and Beason 1976 and Minor et al 1978). Perhaps the most common source of large missiles is collapsing structures. Collapsing structures generate a wide variety of different sizes and types of potential windborne missiles from individual structural members such as joists or purlins, to complete roof and wall sections. It is generally believed that large missiles do not become a major problem until the fastest mile wind speed exceeds 40 m/s.

Figure 8 presents the variation of expected level of impact energy associated with large missile impacts as a function of fastest mile wind speed. The relationship presented in Figure 8 is based on a simulation study of the interaction of hurricanes with a typical residential area (Twisdale et al 1996). This particular simulation study did not produce data for fastest mile wind speeds in excess of 57 m/s.

Over the last several years a major effort has been conducted to require that all windows in both residential and commercial construction be able to resist some level of small and large missile impact criteria. This effort initially involved the presentation of an impact standard that was ultimately adopted as a non-mandatory standard by the Standard Building Code (SBCCI 1994). Impact standards that are similar to the SBCCI impact standard have been adopted by local code authorities. In addition, two recent ASTM standards, E 1886 (ASTM 1997a) and E 1996 (ASTM 1997b), appear to have taken the lead in defining test procedures and specifications for the impact performance of glass. It is felt that these two ASTM standards will ultimately be viewed as the definitive impact criteria adopted by various code groups. The ASTM impact standards include procedures for both small and large missiles impacts followed by cyclic pressure loadings.

It is clear that proper use of either laminated glass or filmed glass can provide reasonable levels of small missile impact resistance. The effectiveness of both types of glass have been demonstrated in both laboratory testing and real-world experience. It is also clear that glass products can be designed to resist various types of large missile impacts. However, it remains to be seen if reasonable glass products can be designed economically to resist the large missile impact criteria contained within the new ASTM standards. Even it is possible to resist the large missile impact levels presented in the new ASTM standards, the level of protection thus provided is not adequate to resist the full effects of a severe hurricane such as Hurricane Andrew that struck Florida in 1992. This judgment is based on the impact severity data presented in Figure 8. In many cases, the most practical solution to the large missile impact problem involves the proper design and use of impact protective devices.

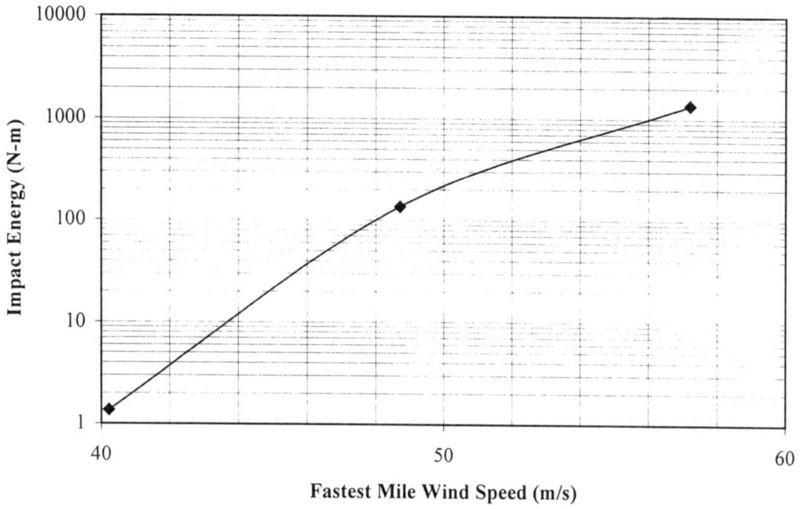

Figure 8 - *Variation of Large Missile Impact Energy vs Wind Speed*

Earthquake Applications

The use of glass in buildings subject to earthquakes has been studied and examined in areas that have experienced the motions, forces and extreme building displacements of these occurrences. It has been observed that the failure of window glass in earthquakes is largely the result of racking forces induced in the glass and glass-to-metal contacts (Lingnell 1994b). Both of these effects are caused by excessive distortions of the frame in which the glass is installed and not the result of lateral loads induced in the glass by the earthquake accelerations. The ability of glass to perform adequately in a building subjected to the effects of an earthquake is dependent on the glass type, thickness, treatment, construction, and framing system.

The objective for the designer is to provide a support system for the glass that will minimize the stresses introduced in the glass through excessive frame distortion and to prevent glass-to-metal contact. The success of the glass in withstanding the effects of an earthquake is dependent on factors that have been identified by the glass industry through years of experience and failure analysis. The factors that must be taken in consideration relate primarily to the framing details. For many years the glass industry has utilized the concept that glass should "float" in the framing. This means that the glass should be properly cushioned and supported during the design event and it should not become a primary element in the support of the other parts of the frame or wall system (Lingnell 1994c). This philosophy provides the core principle for the design of glass to withstand the effects of earthquakes. The glass must be supported in the frame in such a way that it is capable of "riding through the earthquake".

Some of the common framing details that must be controlled include appropriate clearances and proper use of cushioning or shock absorbers. Proper attention to these details will control movements of the framing system so that the glass is not exposed to excessive loads that cause bending, buckling or edge damage to the glass. To effect the proper details, the structural engineer, curtain wall engineer and the architect have to work together using a systems approach to control the amount of movement, displacement, and acceleration that will affect the glass during the event. Designers of the anchorage, glazing pockets or glass attachment methods must consider that the framing system elements will distort, rack, go out of square, and possibly experience permanent deformation while the glass must remain in or return to its original dimensional state.

The proper use of bumper pads, corner protection, setting blocks, and proper attention to adequate edge clearances will assist in avoiding glass-to-metal contact due to frame distortion and allow the glass to ride through the earthquake. It is of utmost importance that the glass edges are protected from impact and possible contact with metal elements in the glazing system such as fasteners, profile protrusions of the aluminum extrusions, setting block chairs, and any other item that can cause edge damage.

The sealants used in a glazing system must also be designed to provide adequate resistance to the anticipated movement that will occur. This is especially true for structurally glazed systems where the retention of the glass unit is dependent on the adhesive and cohesive strengths of the sealant and its ability to elongate without failing. If the sealant fails in a structural glazed system, the glass will be left in an unsupported condition.

Gaskets that are used in a key-in system or roll in wedge type must provide a positive locking method so the gasket cannot disengage from the framing system during the up and down and side to side motions that occur during the seismic event. The intent is to have the gaskets remain in place during the sway and racking that may be encountered during the earthquake. Inter-story drift that occurs during seismic events can cause the glazing system to experience side movements and angular distortion that can dislodge gaskets that are not properly designed or seated during the erection process.

The retention of glass is also extremely important to maintaining the integrity of the building envelope for a structure that has experienced an earthquake. This aspect is vital to keep the elements of weather out of the building as we1l as protecting the contents from vandals. Laminated glass and filmed glass have been shown to enhance the post-breakage performance of the glass subjected to the effects of earthquakes by helping to keep the glass in the frame if it breaks. As with other applications of laminated and filmed glass, the glazing method that is used requires special attention to assure a desirable post-breakage performance. At this point, it is necessary to perform full-scale testing to determine if a particular laminated or filmed glass systems provides the desired post-breakage performance.

Blast Applications

The occurrences of several highly publicized terrorist bombings have focused attention on the design of windows to resist the effects of blasts. While there are several different types of materials that can be used to construct blast resistant windows, this discussion is focused only on the design of windows that incorporate glass.

Thc threat associated with domestic terrorism is relatively new. Therefore, the majority of the structures in use in the United States today were not designed in the context of today's threat. While the potential effects of terrorist acts can be devastating on an individual basis, the frequency of occurrence of such events probably does not warrant a redesign and retrofit of the entire building inventory. However, there is a clear need for the development of methods to reduce the vulnerability of certain high-profile buildings.

The first principle that must be understood in the development of retrofit blast resistant design procedures for glazing is that it is not possible to make a building blast "proof" by simply installing blast resistant windows. For instance, there are no window designs available that would have prevented the collapse of the Federal Building in Oklahoma City. Even if windows were installed that could have resisted the blast pressures, the building system was not capable of resisting the applied blast load. However, it is possible that damage to surrounding buildings could have been reduced through the use of blast resistant glazing. In retrofit situations, the design goal is to develop a glazing system that can resist a blast pressure that is at least as great as the blast pressure that the main structural system of the building can resist. This will assure that the glazing system will not be the first building element to fail.

There are three different glass products that can be incorporated into the design of blast resistant windows: fully tempered glass, laminated glass, and filmed glass. Each of these products can be used to control the effects of blast pressures. However, each of these products has its own unique properties which must be properly understood and exploited if effective blast resistant designs are to be developed.

As discussed previously, fully tempered glass is subjected to a heat-treatment process that locks in a residual surface compression on the surface of the glass. Depending on the level of the residual surface compression that is induced in the glass by the tempering process, the strength of the fully tempered glass can be many times greater than the strength of annealed glass. Thus, fully tempered glass provides an excellent material for improving the blast resistance of windows. In addition, design procedures for the specification and use of fully tempered glass are highly evolved. Properly designed fully tempered glass provides an excellent component for the fabrication of blast resistant windows.

The problem with fully tempered glass is that if the capacity of the glass is exceeded, the glass will fail in a catastrophic fashion. This type of failure can result in injuries to building occupants due to airborne glass particles. However, if the pressure at which the fully tempered glass will fail is larger than the pressure at which the building will collapse, the catastrophic failure of the fully tempered glass becomes a moot point. When it is desirable to improve the post-breakage performance of fully tempered glass, it can be combined with polyvinyl butyral or polyester films to fabricate blast resistant laminated or filmed glass.

As discussed earlier, laminated glass consists of interlaced layers of glass and PVB–based interlayer that are bonded together through the application of heat and pressure. The primary action of the PVB-based interlayer is to act as an adhesive to hold the layers of glass together. Upon failure of the glass layers, the adhesion of the PVB-based interlayer tends to control the release of glass fragments. In addition, if it is properly anchored, the PVB-based interlayer can act as an elastomeric membrane that is capable of resisting a significant pressure. The elastomeric membrane action can be relied on to resist secondary loadings provided that the glass incorporated in the window unit is capable of resisting the primary blast loading.

As discussed previously there is currently considerable controversy surrounding the structural performance of laminated glass. Researchers and manufacturers funded by interlayer manufacturers suggest that laminated glass behaves as though it is monolithic glass if it is loaded at room temperature. However, data collected by researchers not associated with the interlayer manufacturers suggest the temperature of the PVB-based interlayer must be significantly less than room temperature to assure that the behavior approaches that of a monolithic glass plate. (Beason et al 1998)

While an argument can made that it is not likely that the interlayer temperature will exceed 50° C in a windstorm situation, it is completely feasible that the interlayer of a laminated glass unit could be well in excess of 50° C at the time of terrorist action. All researchers who have examined the performance of laminated glass at temperatures in excess of 50° C have recommended that the laminated glass be designed using a layered model for such temperatures. Therefore, it is conservatively recommended that blast resistant laminated glass be designed using a layered structural performance model. This is the best approach unless a detailed finite element analysis can be conducted for the specific problem.

Polyester safety film can be easily applied to the surface of window glass during the fabrication of a window unit or during retrofit operations. Safety film is available in several different thicknesses and tints. It can be affixed to the glass with different types of adhesives. The film can be applied so that it is trimmed flush with the window sight-line in retrofit applications, or the film can be extended to the edge of the glass so that it is fully gripped in the glazing rabbet in new installations. In addition, the film can be mechanically anchored to the window framing system to develop the full strength of the window frame, or it can be anchored to the window frame with a silicone bead.

The performance of a filmed window unit in a blast situation depends on the characteristics of the installation and the properties of the film. While the window film does not increase the pressure at which the glass fails, it can substantially improve the post breakage performance of the glass by containing the glass fragments. In addition, the film is capable of absorbing a significant secondary pressure after the glass has broken. Another important property of window film is that its elastic properties are not as sensitive to normal temperature fluctuations as is the case with the PVB-based interlayer.

Both laminated glass and filmed glass have the potential to offer improved post-breakage performance for glass in blast applications. The mechanisms for accomplishing this improvement is very similar with both products: glass fragments are contained and the film or interlayer provides a post-blast membrane which, if properly designed, can help to keep the glass in the window. In both cases, the key to improved performance is usually the anchorage of the PVB-based interlayer or the PET film to the window frame. To accomplish a successful blast resistant design it is necessary that a complete systems approach be followed where the structural engineer, the curtain wall engineer, and the architect work together to develop a successful design. All elements of the fenestration and surrounding materials must be taken into account. The use of blast resistant laminated glass or filmed glass will provide little benefit if the framing system and its anchorage, as well as the main structural frame of the building are not capable of resisting the blast forces.

The manufacturers of the PVB interlayer have been much more aggressive in performing research for their product than has been the case with the manufacturers of window film. As a result, designers have been provided with an abundance of guidance for the specification of blast resistant laminated glass. Although, it is recommended that great care be taken to assure that any method used for the design of blast resistant laminated glass is based on either a finite element analysis or a layered performance model. While window film manufacturers are substantially behind the PVB manufacturers in conducting full-scale tests, the results of recent blast testing suggests that properly designed and installed window film can provide a performance level that approaches equivalency with laminated glass in some situations.

Based on the above discussion, it is clear that fully tempered glass can be used to significantly improve the blast resistance of windows. In addition, both laminated glass and filmed glass can provide improved post breakage performance of windows. Further, it is clear that additional testing is needed to better define the performance of both laminated and filmed glass in blast situations. Additional research is required to develop reliable design procedures for both laminated and filmed glass. In the interim, it is recommended that blast resistant designs be developed and specified by well informed professionals who are knowledgeable about the properties of both products. Finally, it seems imperative to require independent full-scale testing of all new designs whether they employ laminated or filmed glass.

Conclusions

Over the past several years, the uses and expectations of window glass have undergone dramatic changes. Window glass has evolved from a secondary building element whose use and design was governed by tradition to a primary structural element. The problem with this situation is that the pace of evolution was too brisk for empirically based traditional design procedures to stay abreast of the evolving uses. This situation has led to major misunderstandings in the glass design community. These misunderstandings in turn have lead to the development of conflicting design recommendations. This situation can only be rectified through the introduction of technically defensible analytical design procedures that are based on sound theoretical models, experience, and empirical data. This document presents the reader with a general background concerning the character and properties of various types and constructions of window glass that are currently available. In addition, areas of controversy regarding the behavior of different glass products are discussed. Hopefully, sufficient information is presented to provide direction for an informed designer.

References

Abiassi, J.J., (1981), "The Strength of Weathered Window Glass Using Surface Characteristics", Institute of Disaster Research, Texas Tech University, Lubbock, Tx.

Association of Industrial Metallizers Coaters and Laminators, AIMCAL, November (1996), "Guide to Avoiding Glass Failure".

ASTM, (1997a), "Standard Specification for Heat-Treated Flat Glass-Kind HS, Kind FT Coated and Uncoated Glass", C 1048-97b, Philadelphia, Pa.

ASTM (1997b), "Standard Test Method for Performance of Exterior Windows, Curtain Walls, Doors, and Storm Shutters Impacted by Missile(s) and Exposed to Cyclic Pressure Differentials", E 1886-97, Philadelphia, Pa.

ASTM, (1998), "Standard Practice for Determining Load Resistance of Glass in Buildings", E 1300-98, ASTM, Philadelphia, Pa.

ASTM (1999), "Standard Specification for Performance of Exterior Windows, Glazed Curtain Walls, Doors, and Storm Shutters Impacted by Windborne Debris in Hurricanes", E 1996-99, Philadelphia, Pa.

Beason, W.L., (1980) "A Failure Prediction Model for Window Glass", PhD dissertation, Texas Tech University, Lubbock, Tx.

Beason, W.L., and Morgan, J.R., (1984), "Glass Failure Prediction Model", Journal of Structural Engineering, ASCE, Vol. 110, No. 2, pp 197-212.

Beason, W.L. Meyers, G.E., and James, R.W., (1984), "Hurricane Related Window Glass Damage in Houston", Journal of Structural Engineering, ASCE, Vol. 110, No. 12.

Beason, W.L., (1986a), "Structural Analysis of Sealed Insulating Glass", Journal of Structural Engineering, ASCE, Vol. 112, No. 5, pp 1133-1146.

Beason, W.L., (1986b), "Design Criteria for Insulating Glass With Plates of Equal and Different Thicknesses", Final Report submitted to Cardinal IG, Texas Engineering Experiment Station, College Station, Tx.

Beason, W.L., (1993), "Development of a Failure Prediction Model for Heat-Treated

Glass", Final Report submitted to Cardinal IG, W. Lynn Beason, Ph.D., P.E., Engineering Consultant, College Station, Tx.

Beason, W.L. Kohutek, T.L., and Edel, M.T., July 15 (1998), "Making an Independent Examination of the Structural Behavior of Laminated Glass", Glass Digest, Ashlee Publishing Co., pp 94-98.

Beason, W.L., Kohutek, T.L., and Bracci, J.M., (1998), "Basis for ASTM E 1300 Annealed Glass Thickness Selection Charts", Journal of Structural Engineering, ASCE, Vol. 124, No. 2, pp 215-221.

Beason, W.L. and Norville, S.H., (1989), "Development of a New Glass Thickness Selection Procedure", Proceedings 6th U.S. National Conference on Wind Engineering, Vol. II, University of Houston, Houston, Tx.

Behr, R.A., Minor, J.E., and Norville, H.S., (1993), "Structural Behavior of Architectural Laminated Glass", Journal of Structural Engineering, Vol. 119, No. 1, pp. 202-222.

Fariss, R.H., (1993), "Fifty years of safer windshields," Chemtech, Ameri. Chem. Soc., Vol. 23, No. 9, pp. 38-43.

Glass Association of North America, (1997), "GANA Glazing Manual, 1997 Edition", , Topeka, Ks.

Hershey, R.L. and Higgins, T.H., (1973), "Statistical Prediction Model for Glass Breakage from Nominal Sonic Boom Loads", Report No. FAA-RD-73-79, Booz-Allen Applied Research, Inc., Bethesda, Md.

Hooper, J.A., (1973), "On the Bending of Architectural Laminate Glass", International Journal of Mechanical Science (Great Britain), Vol 15, pp. 309-323.

Lingnell, A. W., March (1981), "Energy Saving Designs with Glass", Proceedings of Symposiums Sponsored by the EITEL Institute for Silicate Research - The University of Toledo and the American Ceramic Society.

Lingnell, A. W., Oct. (1994a), "Glass Deflection Characteristics for Wall System Design", U. S. Glass, pp. 54-55.

Lingnell, A. W., Feb. (1994b), "Glass and Glazing Relative to Curtainwall Design", U. S. Glass, pp. 40-47.

Lingnell, A. W., March (1994c), "Initial Survey and Audit of Glass and Glazing System Performance During the Earthquake in the Los Angeles Area on January 17, 1994", Final Report submitted to Primary Glass Manufacturers Council, Lingnell Consulting Services.

Mencik, J., (1992), Strength and Fracture of Glass and Ceramics, Elsevier, Amsterdam.

Minor, J.E. and Beason, W.L., (1976), "Window Glass Failures in Windstorms", Journal of the Structural Division, ASCE, Vol. 102, No. ST1, Proceedings Paper 11834, pp. 147-160.

Minor, J.E., Beason, W.L., and Harris, P.L., (1978), "Designing for Windborne Missiles in Urban Areas", Journal of Structural Division, ASCE, Vol. 104, No. ST11, Proceedings Paper 14143, pp. 1749-60.

Norville, S.H., and Minor, J.E., (1985), "Strength of Weathered Window Glass", American Ceramics Society Bulletin, Vol. 64, No. 11, pp 1467-1470.

Southern Building Code Congress International, Inc., (1994), "SBCCI Test Standard for Determining the Impact Resistance From Windborne Debris", SSTD 12-94, Birmingham, Alabama.

Schrader, B.D., (1982), "The Effects of Age, Surface Damage, and Humidity on the Strength of Window Glass", ME report, Texas A&M University, College Station, Tx.

Twisdale, L.A., Vickery, P.J., and Steckley, A.C., March (1996), "Analysis of Hurricane Windborne Debris Impact Risk for Residential Structures", Applied Research Associates, Inc., Raliegh, North Carolina.

Index